A GUIDE TO FRAMEWORK MOLECULAR MODELING

DARLING MODELS INC.

Stephen D. Darling

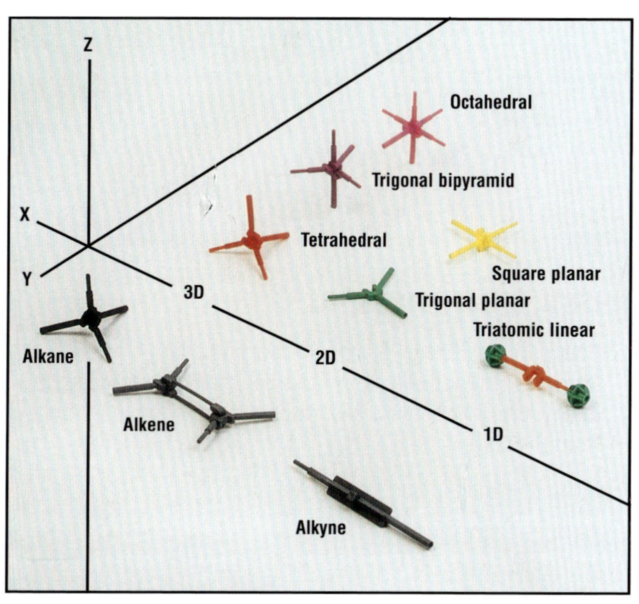

ISBN 978-0-9648837-03

Cover:

The axis is to remind the reader that all atoms are three dimensional, but that bonds made with each atom may result in three dimensional (3D), two dimensional (2D), and one dimensional (1D) regions in a molecule. It is the exploration of these domains where tactile models are most useful. The models predict intramolecular relationships; the basis of conformational analysis, and illustrate stereochemistry.

Special thanks to Professor Cal Y. Meyers,
Southern Illinois University, Carbondale

ISBN 978-0-9648837-03

Printed in the United States of America

DARLING MODELS INC.
P. O. BOX 1818
STOW, OH 44224

VOICE: 330-688-2080
FAX: 330-688-5750
E-MAIL: darling@darlingmodels.com
WEB SITE: www.molecularvisions.com
or: www.darlingmodels.com

CONTENTS

Introduction 7

Section

1: Relating models to two dimensional drawings 7
- Water 7
- Hydronium ion, ammonia, carbanion 8
- Ethane, ethene (ethylene), ethyne (acetylene) 9
- Methanol (methyl alcohol), ethanol (ethyl alcohol), dimethyl ether 9
- Carboxylate ion, nitrate ion, 10

2: The MOLECULAR VISIONS™ model pieces 10
- Common 2nd row elements 10
- 3rd row and beyond: trigonal and octahedral pieces 11
- Bonding pieces 12
- Space filling ATOM VISIONS™ balls 13
- Color 14

3 : Assembly and disassembly of atoms "with bonds" 15
- Tetrahedral atom of carbon, oxygen, and nitrogen 15
- Atoms joined by a pi system (double bond and triple bond): 16
  - A. alkenyl (olefinic), carbonyl, imino or immonium, and cumulenyl groups 16
  - B. aromatic and nonaromatic systems 18
  - C. triple-bonded atoms 18
- The marker balls 18
- Linear, trigonal, trigonal-bipyramid and square-planar atoms 19
- Octahedral and square-pyramid atoms 20
- Lengthening bonds 21
- Inserting framework atoms "with bonds" into an ATOM VISIONS™ shell 21

4: Using the atom model "with bonds" to create molecules 22
- Three-membered rings and other strained systems 23
- Projections: Acyclic and cyclic molecules 24
  - A. wedge and dotted-line projections 24
  - B. Fisher projections 24
  - C. Newman projections 25
  - D. sawhorse projections 26

5: Stereoisomers 29
- Configurational-Geometric isomers 29
  - A. cyclic compounds 29
  - B. carbon-carbon double-bonded compounds 31
  - C. metal centers 31
- Chiral isomers: enantiomers, optical isomerism 32
  - A. simple aliphatic compounds 32
  - B. metal centers 33
  - C. chiral molecules without sp^3 chiral atoms 34

6: Models to investigate reactions 35
- Reactions 35
- Retrosynthesis 36
- Using models to interpret spectra 37

CONTENTS

7: Model building for enhanced visualization 39

  ATOM VISIONS™ hemispheres to enlarge the atom center 39

  Use of colored marker balls in modeling 40

    A. markers for differentiating "like" atoms 40

    B. orbital symmetry 40

      1. electrocyclic stereochemistry 41

      2. sigmatropic rearrangements 42

  Use of "nontraditional" colors and two-toned atom models 42

    A. markers 42

    B. conformations 43

8: Resonance models 43

9: Metal ligands 44

10: Hydrogen bonds 46

TABLE OF FIGURES

1. Water represented as A, condensed formula, B, ball and stick, C, Lewis dot formula 7
2. Molecular Visions™ modeling of water 8
3. A, hydronium ion, B, ammonia, C, carbanion 9
4. Models constructed from molecular formulas A, ethane C_2H_6, B, ethene C_2H_4, C, ethyne C_2H_2 9
5. Models of CH_4O (methanol) and C_2H_6O (ethanol and dimethyl ether) 10
6. Models of carboxylate and nitrate anions 10
7. The common 2nd row element representations 11
8. Other Molecular Visions™ representations of atoms 11
9. Bonding pieces used in coordination 12
10. ATOM VISIONS™ joined tetrahedral hemispheres and their respective balls around tetrahedral atoms "with bonds" 13
11. ATOM VISIONS™ joined trigonal-octahedral hemispheres and their respective balls around trigonal and octahedral atoms "with bonds" 13
12. Sp^3 pieces, trigonal pieces and marker balls in many colors 14
13. Model representations of the tetrahedral atom "with bonds" 15
14. Assembling the tetrahedral atom, "with bonds" 15
15. Disassembling the tetrahedral atom 16
16. Various ways to form an alkenyl, a carbonyl, an imino or immonium, and cumulenyl groups 16
17. The assembly of doubly-bonded atoms 17
18. Model representations of benzene, pyridine, pyrrol, furane, and naphthalene 18

19. Model representations of cyclobutadiene, cyclooctatetraene, and cyclodecapentaene 18
20. Marker balls used to represent atoms or groups, electron pairs, hydrogen bonds, and "p" orbital symmetry 19
21. Atom centers, "with bonds" from combinations of linear and trigonal pieces 19
22. Assembly and disassembly of a trigonal atom 20
23. Atom centers incorporating the octahedral piece 20
24. Assembly and disassembly of an octahedral atom center, "with bonds" 21
25. The tetrahedral ATOM VISIONS™ hemispheres being placed over the framework atom "with bonds" 21
26. The trigonal-octahedral ATOM VISIONS™ hemispheres being placed over the framework atom "with bonds" 22
27. The trigonal-octahedral ATOM VISIONS™ hemispheres placed over the framework atom "with bonds" pi system of the double bond 22
28. Joining two tetrahedral atoms by inserting the rod into the tube 22
29. Forming a three-atom ring 23
30. Models of cyclopropane, an epoxide, and an aziridine 23
31. Methane represented as a wedge and dotted-line drawing, A, or Fisher projection, B 24
32. Fisher projections of CH_3CH_2OH (ethyl alcohol), A, and $HOCH_2CHOHCHO$ (D-glyceraldehyde), B 24
33. Modeling a Fisher projection. A. $HOCH_2CHO$,
B. $HOCH_2CHOHCHO$ (D-glyceraldehyde),
C. $HOCH_2[CH(OH)]_2CHO$,
D. $HOCH_2[CH(OH)]_4CHO$ (D-glucose) 25
34. D-glucopyranose 25
35. Newman projections of the conformations of 1-propanol 25
36. Model representations of the conformations of 1-propanol looking down the C1-C2 bond 26
37. The sawhorse projections of the 1-propanol conformations 26
38. Molecular model representations of the sawhorse projections 27
39. Newman projections of the boat and chair conformations of cyclohexane 27
40. Model representations of the views shown in Figure 39 28
41. Sawhorse projections of the boat and chair conformations of cyclohexane 28
42. 1,2 substitutions on cyclohexane, cis and trans isomers 30
43. Representation of cyclohexane on a plane 30
44. Nonbonded methyl/hydrogen interactions between axial groups on a methylcyclohexane 30

45. Geometric isomers of 2-amino-3-methoxy-2-butene 31
46. Geometric isomers of the square-planar platinum dichloride ammonia complex 31
47. Geometric octahedral isomers 32
48. Drawings to define R and S configuration at a chiral center 33
49. Optical isomers of a tris-ethylenediamine complex 33
50. A chiral R *trans*-cyclooctene 34
51. Chiral allenes 34
52. A reaction; Bromination of an *E*-alkene 35
53. Models of a protonated alkene; carbocation intermediates 35
54. Ozonolysis of an alkene 36
55. An example of modeling a retrosynthesis: "Dissection" of the preparation of the tertiary alcohol 2,3-dimethyl-3-pentanol 37
56. Correct and incorrect disconnections in a retrosynthesis 37
57. Molecular fragments determined from spectral interpretations 38
58. ATOM VISIONS™ enhanced, 1)(1S,3R)-3-methylcyclohexanol-1. 2) Glucose 39
59. ATOM VISIONS™ enhanced, 1) Glycine, 2) Toluene 39
60. Colored marker balls used to mark "equivalent" and "non-equivalent" hydrogens 40
61. Models of orbital symmetry 40
62. Modeling aromatic systems 41
63. Use of color-coded models to determine the course of an electrocyclic reaction 41
64. Modeling a sigmatropic rearrangement 42
65. Marking bridgeheads with colored atoms 42
66. Models showing the use of two-colored atoms 43
67. Construction of a resonance hybrid model of an amide 43
68. Model of a pi bonded ligand 44
69. Modeling organometallic ligands. A. A benzene ligand; B. A tetrahapto ligand; C. An allyl ligand; D. A cyclopentadienyl ligand; E. A biscobalt alkyne complex 44
70. A model of ferrocene 45
71. A porphine ligand 45
72. Modeling hydrogen bonding involving H_2O 46
73. A base pair 46
74. Protein Alpha-helix 47
75. A Molecular Visions™ model of an ice crystal lattice 48

SPECIAL NOTE — CARING FOR THE BONDING PIECES IN THIS KIT

The material used for the manufacture of these models is a thermoplastic polypropylene copolymer. It will bend when sufficient pressure is applied. Treatment with boiling water for one minute will return the pieces to their original angles, restore the dimensions of pieces that were cold formed, and remove distortions in pieces which were kept for an extended period in models of strained molecules.

The pieces should be grasped firmly when pushing the rod into the tube so as not to bend or shear the rod piece. Occasionally the joining offers more friction than desired. This may be alleviated with a light spray of silicone or a thin film of Lubriplate® lubricant. Oil lubricants do not seem to work as well.

INTRODUCTION

Molecular models are as vital a tool for the study of chemistry as a calculator is the study of mathematics. The purpose of this text is to provide examples of how models and model pieces may be utilized. Molecular Visions™ models may be assembled into infinite combinations so the user can construct not only familiar configurations but can also explore undiscovered possibilities. The examples illustrated in this text are not meant to place limitations on their use.

As with all tools, the more models are used the better they will serve the user. Models are intended to inspire the imagination, stimulate thought and assist the visualization process. They present the user with a solid form of an abstract object that is otherwise only formulated in a chemist's mind, speech or in written text. Chemistry textbooks contain a pictorial language for describing molecules and reactions, however, molecular models enhance comprehension through more vivid association.

The scale of the models is: 2.0 in. (50.8 mm) = 100pm (1.00Å).

RELATING MODELS TO TWO-DIMENSIONAL DRAWINGS

WATER

Most textbooks and many instructor notes contain drawings made up of letter symbols and lines to represent the atoms and bonds of molecules. In some books the water molecule (H_2O) is represented in various ways, as shown in Figure 1.

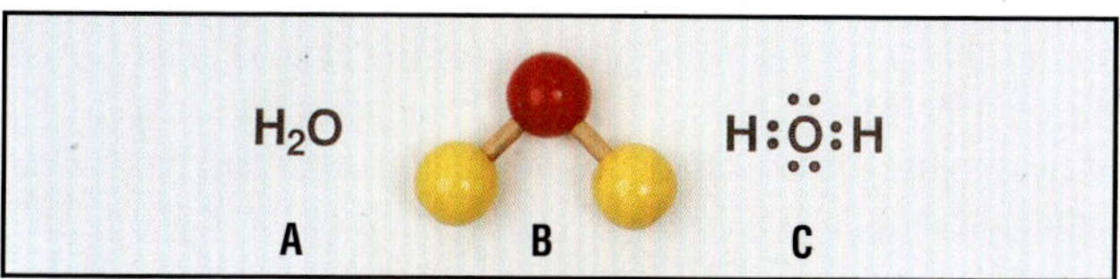

Figure 1. *Water represented as A, condensed formula, B, ball and stick, C, Lewis dot formula*

All of the representations in Figure 1 lack some important feature about the molecule of water. In A, neither the two nonbonded pairs of electrons nor bond angles are illustrated; in B, the nonbonded electron pairs are missing; and in C, the bond angle shown is incorrect.

Water may be modeled to show that the two hydrogen atoms are bonded to oxygen and not to each other (i.e. the H_2 is not H-H). Remember, modeling is a tool to help the mind. The user may use what ever is available to this end. The color red has been chosen to represent oxygen, because that color has been accepted by The International Union of Crystallography. (Another color could be used to represent the oxygen if that is the only tetrahedral atom available.) The important thing here is the shape of the molecule. Accuracy of bond lengths and angles are not critical in modeling here, as small differences will not be noticed. Figure 2, shows some possible ways water may be modeled.

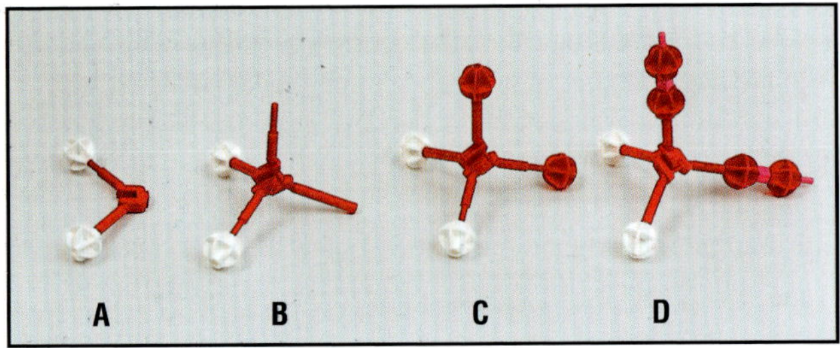

Figure 2. *Molecular Visions™ modeling of water, A, divalent oxygen atom with bonds to two hydrogens (white ball), no electron pairs, B, oxygen atom with tetrahedral bonds, two bonds are to hydrogen atoms and two bonds each representing an electron pair, on oxygen; C, same as B, but each electron pair of oxygen being represented by a red ball; D, same as B, but each electron pair on oxygen being represented by two red balls*

HYDRONIUM ION, AMMONIA, CARBANION

With models the structural similarities between many compounds with different atoms become apparent. The model also represents an accounting of the electrons. A covalent bond contains two electrons. Each atom is shown with four bonds; therefore, there are eight "valence" electrons around each atom. Electrons between atom centers are shared. When counting bonding electrons only one electron is counted for each atom. There are only five bonding electrons on the oxygen atom in the hydronium ion; for a neutral oxygen atom there should be six. Thus, we write a positive charge (+) on the oxygen atom. The ammonia molecule has the same shape as the hydronium ion, but no charge. In the model of ammonia the pink ball represents the electron pair on nitrogen. The carbanion has the same shape as hydronium ion and ammonia, but it has a negative charge by a similar accounting of electrons. Figure 3, shows the shape of the hydronium ion (H_3O^+), ammonia (NH_3 or H_3N), and the carbanion ($C\,H_3^-$ or $H_3\,C^-$).

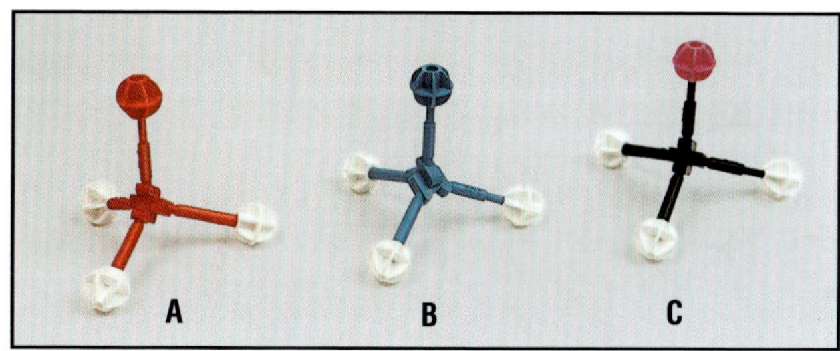

Figure 3. *A, hydronium ion, B, ammonia, C, carbanion*

ETHANE, ETHENE (ETHYLENE), ETHYNE (ACETYLENE)

Alkanes (paraffins), alkenes (olefins), and alkynes (acetylenes) are often discussed early in chemistry texts. Examples are ethane (C_2H_6), ethene (ethylene, C_2H_4) and ethyne (acetylene, C_2H_2). Successful modeling depends on selecting atom centers with the correct number of sigma bonds to satisfy all the bonds to each atom. All electrons must be accounted for. They must be counted as bonded- or nonbonded electron pairs. Hydrogen atoms may only form one bond (sigma). Therefore, the two carbon atoms in these examples must be joined (bonded) for the structure to hold together. The choices in these examples, respectively, are two carbons joined by a single bond, two carbons joined by a double bond, and two carbons joined by a triple bond. Figure 4 shows how these may be modeled.

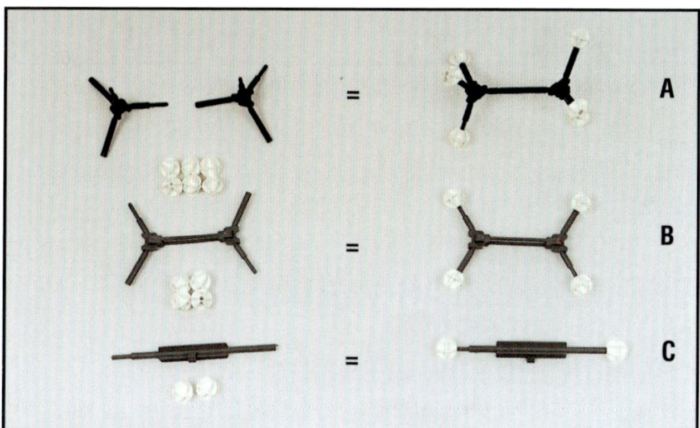

Figure 4. *Models constructed from molecular formulas A, ethane,* C_2H_6*; B, ethene,* C_2H_4*; C, ethyne,* C_2H_2

METHANOL (METHYL ALCOHOL), ETHANOL (ETHYL ALCOHOL), AND DIMETHYL ETHER

Molecular modeling will easily illustrate that the molecular formula CH_4O represents only one compound, but the molecular formula C_2H_6O, represents two compounds, as illustrated in Figure 5.

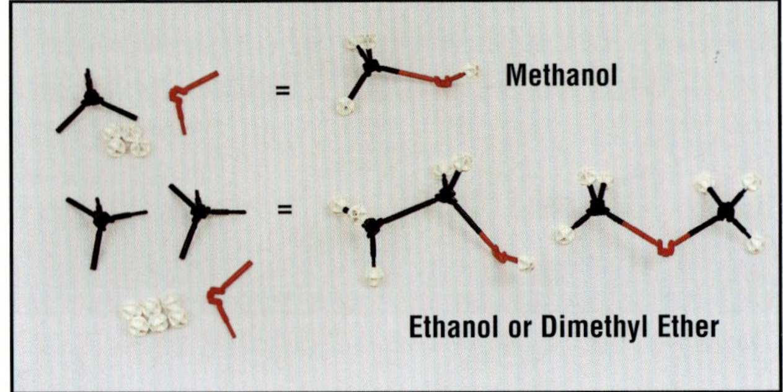

Figure 5. *Models of CH_4O (methanol) and C_2H_6O(ethanol and dimethyl ether)*

CARBOXYLATE ION AND NITRATE

The R-CO_2^- and NO_3^- anions are referred to as resonance stabilized: more than one structure may be drawn to illustrate them, as demonstrated with the models shown in Figure 6.

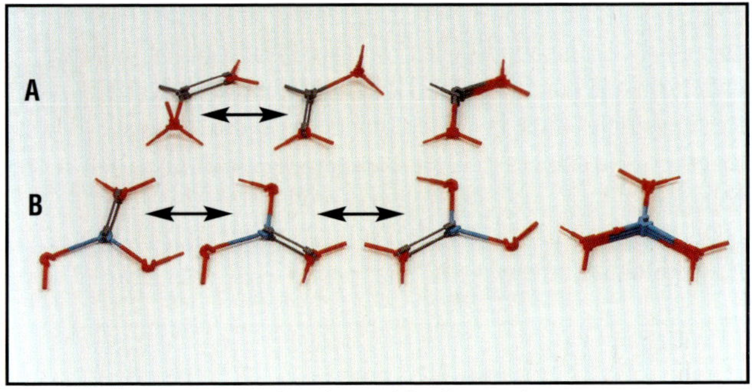

Figure 6. *Models of A, carboxylate and B, nitrate anions*

The carboxylate has two valid forms and the nitrate three forms. The additional structures at the far right are model representations of the resonance forms.

THE MOLECULAR VISIONS™ MODEL PIECES

COMMON 2ND ROW ELEMENTS

On page 11, are pictures of the tools we shall be using for our modeling. The pieces in Figure 7 are the most commonly used in modeling organic compounds and many inorganic compounds containing 2nd-row elements.

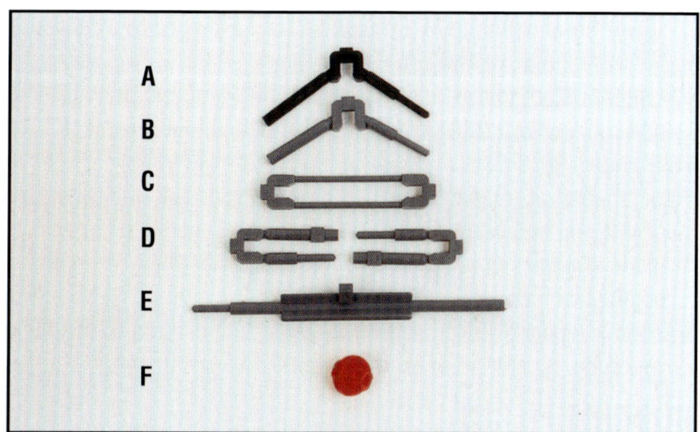

Figure 7. *The common 2nd row element representations.*

1) tetrahedral-atom piece, A, made up of a rod and a tube attached to a central "U" at an angle of 109^0 (referred to hereafter as the sp^3 piece).
2) trigonal atom piece, B, made up of a rod and a tube attached to a central "U" at an angle of 120^0 (referred to hereafter as the sp^2 piece).
3) double-bond-pi piece, C. represent only the pi bond and do not include the sigma bond or atom markers.
4) half-double-bond-pi pieces, D, when joined may replace the double-bond piece. The cube towards the end is used in attaching a pi-bonding piece.
5) linear triple-bonded carbon-carbon pairs, E. The cube in the center of one pi bond is used in attaching a pi bonding piece. Unlike #3 and #4 this piece represents two sp hybridized atoms joined by a triple bond.
6) non-rolling-marker ball, F.

3RD-ROW AND BEYOND: TRIGONAL AND OCTAHEDRAL

Figure 8, presents the model parts most commonly used to depict atom hybridization of second-row elements or coordination compounds of third-row elements and beyond.

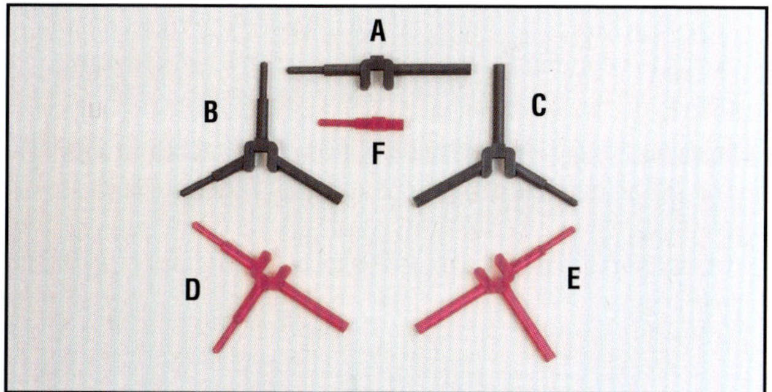

Figure 8. *Other Molecular Visions™ representations of atoms.*

1) linear bond made up of a rod and a tube attached to a central "U" at an angle of 180°, A.
2) trigonal atom pieces made up of two rods and one tube, B, or one rod and two tubes coplanar, C, attached to a central "U" at angles of 120°.
3) octahedral-atom pieces, made up of two rods and one tube, D, or one rod and two tubes, E, attached to a central "U" in a plane at angles of 90° (referred to hereafter as the, octahedral piece).
4) bond extender a short linear piece consisting of a rod connected to a tube, representing an increment of 44 pm (0.44Å), F.

BONDING PIECES

What we call a "bond" is a mnemonic to visualize and rationalize the link connecting atoms and, therefore, the shapes of molecules. For this reason we have supplied pieces which do not represent atom parts but permit the atoms and groups to be connected, allowing us to "visualize" complex molecules. Figure 9 illustrates these bonding pieces.

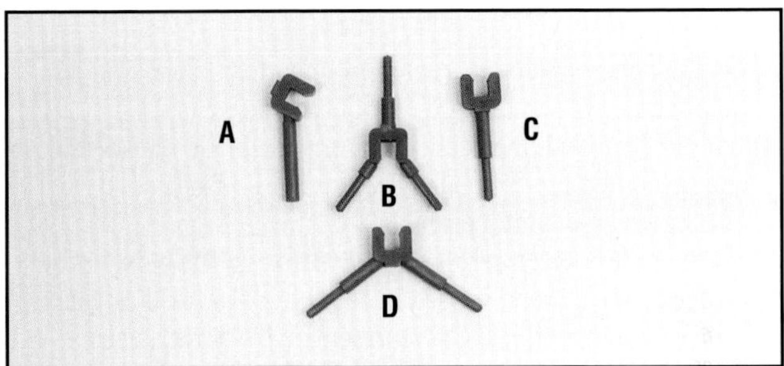

Figure 9. *Bonding pieces used for coordination*

1) a 60° bonding piece, A, used as a pi coordination bonding piece to form hapto bonds and allylic pi bonds.
2) cyclopentadienyl hub, B made up of the central "U" with three rods to form the center of a cyclopentadienyl ring for attach ing a metal. Also used in forming tetrahapto bonded metals.
3) a pi bonding piece, C, consisting of a "U" center with a rod or tube attached. It is used to attach atoms to the cubes on pi bonds to bond metals and form three-center two-electron bonds.
4) reverse sp^2 piece, D, made up of two rods attached to a central "U" at an angle of 120° but at the end opposite of the sp^2 piece. It may be used to form cobalt-alkyne complexes or, when used with a trigonal piece, a distorted trigonal bipyramid.

SPACE FILLING ATOM VISIONS™ BALLS

Figure 10. *ATOM VISIONS™ joined tetrahedral hemispheres and their respective balls around tetrahedral atoms "with bonds".*

Figure 11. *ATOM VISIONS™ joined trigonal-octahedral hemispheres and their respective balls around trigonal and octahedral atoms "with bonds".*

COLOR

Single atoms of elements do not have color; only as bulk solids (and some liquids) do they exhibit color. As stated earlier, the color red was chosen for pieces representing oxygen because of an international agreement which has selected certain colors to represent common atoms. Liquid oxygen is, in fact, blue. A list of these color codes follows:

white	Hydrogen	green	Bromine
black	Carbon	dark green	Iodine
blue	Nitrogen	black	Silicon
red	Oxygen	purple	Phosphorus
yellow green	Fluorine	yellow	Sulfur
light green	Chlorine	hot pink	undesignated atoms

The color codes are generally used, unless other color codes are clearly defined by the user. Confusion in communicating should be avoided. Color preference is a personal choice; the selection is limited only by what is available. The shape of the model and its usefulness to the user is most important, not color. At least one user of these models is known to prefer purple. In the later section suggestions are presented on how to use color to emphasize certain aspects of a model or as a reminder to the viewer. Figure 12, illustrates the variety of colors available, but the piece shape remains the same.

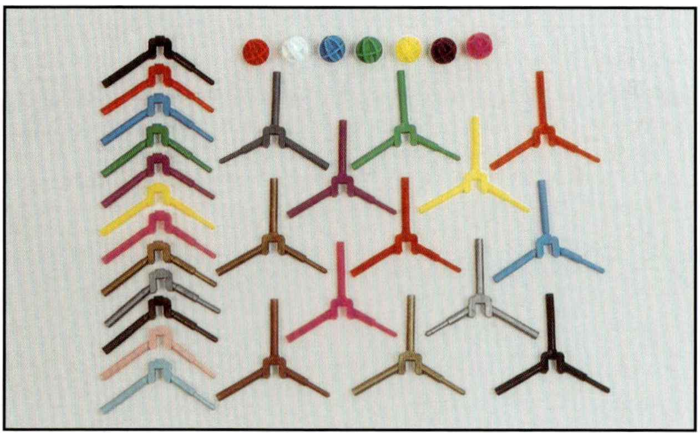

Figure 12. *sp^3 pieces, trigonal pieces, and marker balls in many colors*

ASSEMBLY AND DISASSEMBLY OF ATOMS "WITH BONDS"

TETRAHEDRAL ATOM OF CARBON, OXYGEN, AND NITROGEN

Figure 13, illustrates the use of sp^3 pieces to form the atom centers of carbon, oxygen and nitrogen "with bonds".

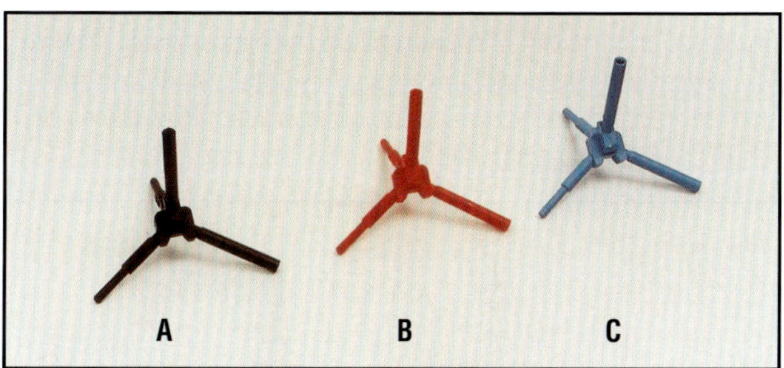

Figure 13. *Model representations of the tetrahedral atom "with bonds"*

1) **The tetrahedral carbon** "with bonds". The black sp^3 pieces are always used in joined pairs to represent the four bonds of a tetrahedral sp^3 hybridized carbon atom, A.
2) **The oxygen atom** "with bonds". The red sp^3 pieces may be used alone to represent the two bonds of oxygen; or joined to form the two bonds of oxygen and its two pairs of nonbonded electrons, B.
3) **The nitrogen atom** "with bonds". The blue sp^3 pieces are always used in joined pairs to represent the three bonds and one lone pair of nonbonded electrons on nitrogen, C.

The assembly of a tetrahedral atom from two sp^3 pieces is carried out as illustrated in Figure 14.

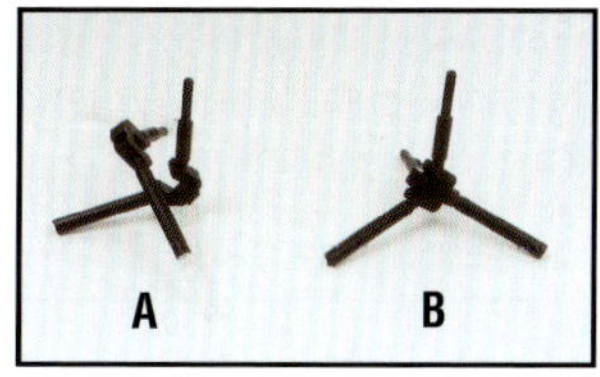

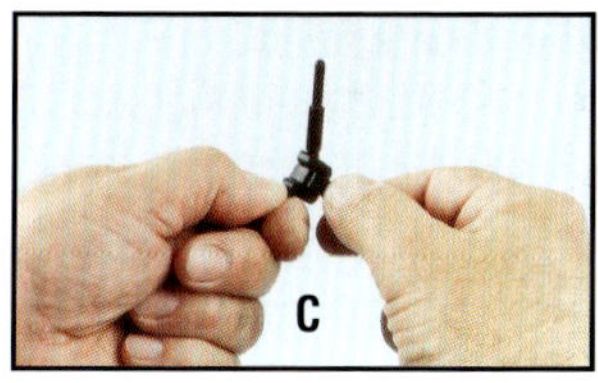

Figure 14. *Assembling the tetrahedral atom, "with bonds"*

1) slide the "U" openings together at right angles, A.

2) pinch the two pieces together until they click, B.
3) grasp the two pieces against the central "U". Pull sharply with the left hand and push sharply with the right hand until there is a second click, C.

The tetrahedron may be taken apart by spreading the "V" shaped bonds on one piece, to unlock the teeth, while pushing it out of its locked position. This may be accomplished in one motion with one hand, by placing two or four fingers across the "V" of one piece and the ball of the thumb on the opposite side (Figure 15). A gentle squeeze will spread the "V" slightly and push the two pieces apart. In Figure 15, the left hand stabilizes the piece while the right hand spreads both pieces and separates them.

Figure 15. *Disassembling the tetrahedral atom*

ATOMS JOINED BY A PI SYSTEM (DOUBLE BOND AND TRIPLE BOND)

A. Alkenyl (olefinic), carbonyl, imino or immonium, and cumulenyl groups

The gray sp^2 piece, double-bond-pi piece, and half-double-bond-pi piece must always be used together or with some other piece; they do not stand alone. Figure 16, illustrates the use of sp^2 pieces, other atom markers and double-bond-pi piece and/or half-double bond-pi piece to make an alkenyl, carbonyl, imino or immonium, and cumulenyl groups.

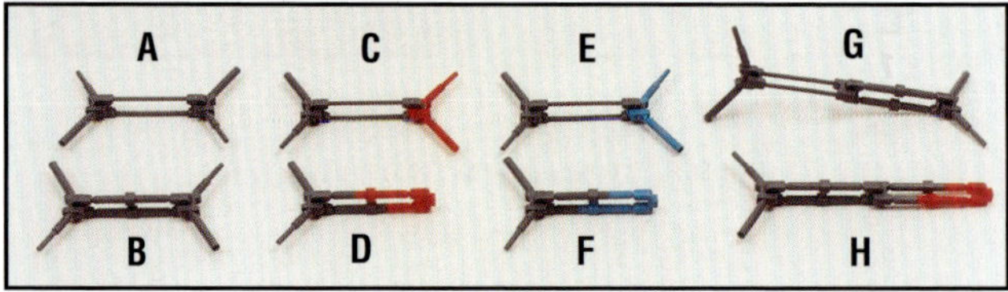

Figure 16. *Various ways to form an alkenyl, a carbonyl, an imino or immonium, and cumulenyl groups*

1) **Carbon-carbon double bond (Alkene) C=C.** A gray double-bond-pi piece is connected to two sp^2 gray pieces, A . Two gray half-double-bond-pi pieces are joined and connected to two gray sp^2 pieces, B.
2) **Carbon-oxygen double bond (Carbonyl) C=O.** A gray double-bond-pi piece is connected to one sp^2 gray piece and one red piece. (Depending on the kit you have, you may use a sp^3 or sp^2 red piece), C. A gray half-double-bond-pi piece and a red half-double-bond-pi piece are joined and a gray sp^2 piece connected to the gray half-double-bond-pi piece, D.
3) **Carbon-nitrogen double bond (imino or immonium) C=N.** A gray double-bond-pi piece is connected to one sp^2 gray piece and one blue piece. (Depending on the kit you have, you may use a sp^3 or sp^2 blue piece.), E. A gray half-double bond-pi piece and a blue half-double-bond-pi piece are joined and a gray sp^2 piece connected to the gray half-double bond-pi piece, F).
4) **Cumulenes (allenic and ketenic) C=C=C and C=C=O.** The allene system may be constructed by inserting a half-double-bond-pi piece into a double-bond-pi piece joined to another half-double-bond-pi piece. The sp^2 pieces are inserted into the remaining ends of each double bond, G. The ketene system may be constructed when a half-double-bond-pi piece is inserted into a double-bond-pi piece and joined to another red half-double-bond-pi piece. An sp^2 piece is inserted into the remaining end of the carbon-carbon double bond, H.

The assembly of a pi system with atom markers and bonds is carried out by joining a pi double bond piece with an atom piece such as the gray sp^2 carbon marker, as shown in Figure 17.

***Figure 17**. The assembly of doubly-bonded atoms*

The pi piece is being pushed down into the sp^2 piece with the fingers and the sp^2 piece is pushed up with the thumbs. Later the assembly may be carried out in one step similar to that shown in Figure 14C. The second atom marker should be added with the tube and rod on the opposite side from the first marker to avoid polarity problems later. Disassembly again requires the atom marker bonds to be spread slightly and the pi system pushed through.

B. Aromatic and nonaromatic systems

The double bond systems created in Figure 16 may be joined to represent other interesting multiple double bond systems. The aromatic and nonaromatic pi systems may be represented by a combination of double bonds to form rings. Some of these systems are shown in Figures 18 and 19.

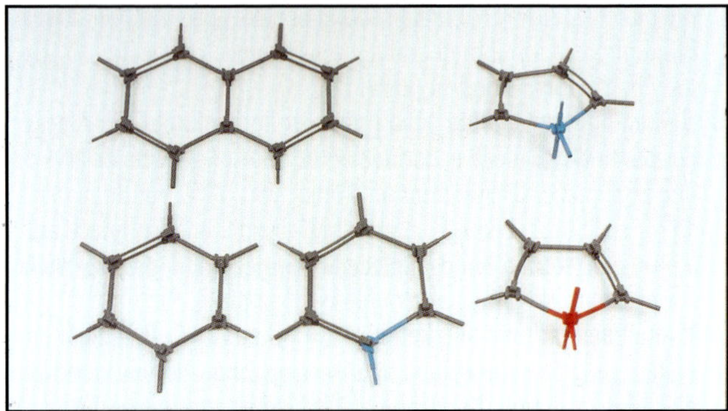

Figure 18. *Model representations of (front, L-R) benzene, pyridine, furane, (back, L-R) naphthalene and pyrrol.*

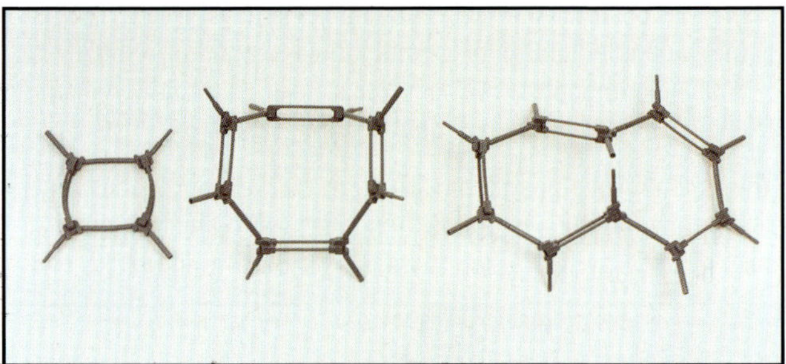

Figure 19. *Model representations of (L-R) cyclobutadiene, cyclooctatetraene, and cyclodecapentaene.*

C. Triple-bonded atoms

Triple-bonded atoms are represented by one piece already pictured in Figure 7E. The gray piece represents the sigma bond and two pi bonds (the fins) between two atoms. Extending beyond the atoms are the two sp sigma bonds to which other atoms are joined. The group may also represent the carbon-nitrogen triple bond of a nitrile(C≡N) or the carbon-oxygen triple bond (C≡O) of some carbon monoxide derivatives.

THE "MARKER BALLS"

The "marker balls" are the wild cards. They may be used to represent anything, e.g. atoms, groups etc. Thus they may be used with bond extenders to form electron pairs, hydrogen bonds, and "p" orbital symmetry markers for rearrangements. Examples of these uses are shown in Figure 20.

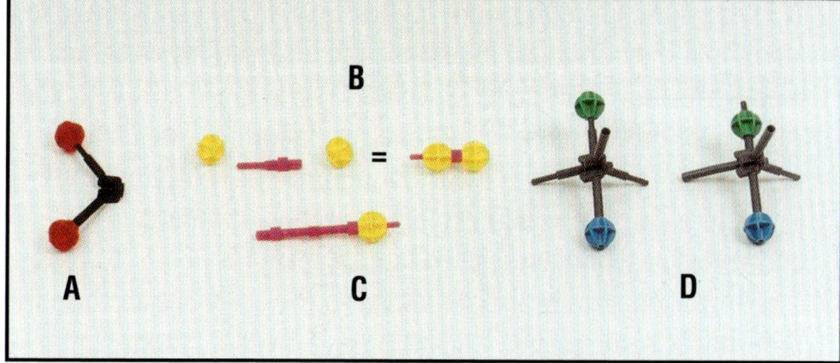

Figure 20. *Marker balls to represent atoms, groups, electron pairs, hydrogen bonds, and "p" orbital symmetry.*

1) The small hole of the marker ball fits snugly at the tip of the rod end so that the ball is the same distance from the atom center as a ball put on the tube of the same piece, A.
2) An electron pair marker may be made from a bond extender and two marker balls. The large hole of each ball fits over the bond extender; the ball is pushed up to the cube leaving a rod protruding. Either end of the piece maybe attached to a bond from an atom, B.
3) The hydrogen bond for base pairs and proteins may be made with three bond extenders and one marker ball. The rod end protruding through the ball is inserted into the tube of the donor atom, C.
4) Marker balls may be used to denote the orbital symmetry of "p" orbitals (described later under "sigmatropic reactions"). Placing the large hole of the ball onto the rod end of a bond permits the rod to protrude through the ball so as to be used in bonding to another tube piece, D.

LINEAR, TRIGONAL, TRIGONAL-BIPYRAMID, AND SQUARE-PLANAR ATOMS

The linear bond represents bivalent atoms such as beryllium or mercury. The trigonal bond piece represents trivalent atoms such as boron. Combinations of these pieces with other pieces may be used to represent a variety of other atom centers "with bonds". Many examples are shown in Figure 21.

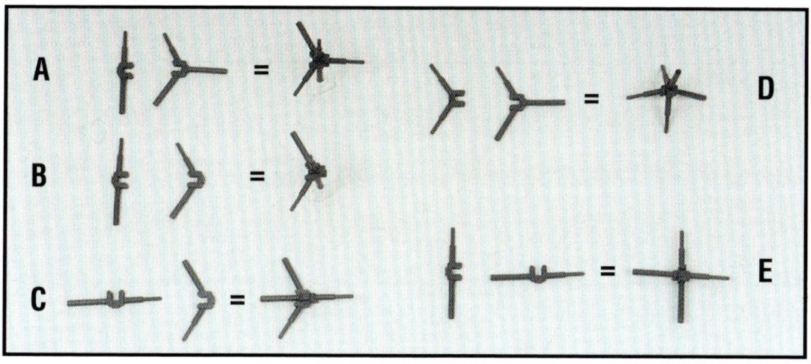

Figure 21. *Atom centers, "with bonds" from combinations of linear and trigonal pieces*

1) A linear bond joined at right angles with a trigonal atom gives a trigonal bipyramid, A. Assembly and disassembly of these pieces are show in Figure 22.
2) A trigonal bipyramid missing one sp^2 ligand may be formed from an sp^2 piece and a linear bond joined at right angles, B. This piece is useful when exploring orbital-symmetry-controlled reactions described later, Figure 63.
3) The linear bond may also be joined to the sp^2 piece so they are both in the same plane. This becomes a trigonal atom center with an rod or tube attachment for anchoring the atom to a metal ligand, C. Its use is described later, Figure 69.
4) A distorted trigonal bipyramid is formed when a reverse sp^2 piece is joined to a trigonal atom, D.
5) Joining two linear bonds gives a square planar atom center, "with bonds", E.

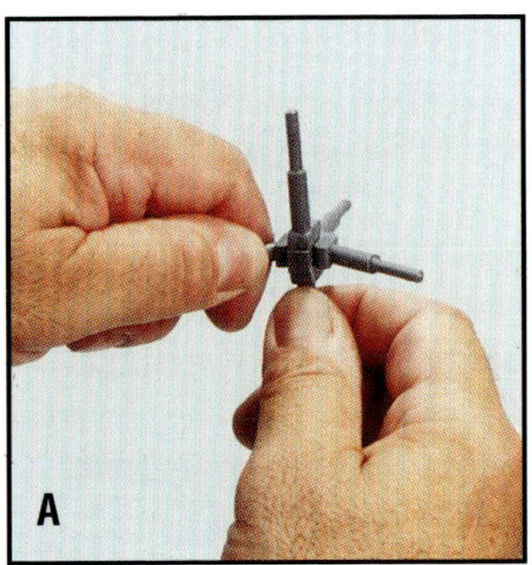

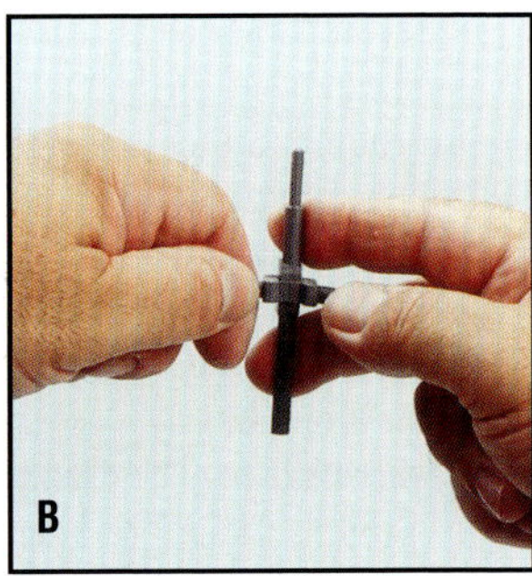

Figure 22. *A) Assembly and B) disassembly of a trigonal bipyramid*

OCTAHEDRAL AND SQUARE-PYRAMID ATOMS

The pink octahedral atom piece represents half of an atom with octahedral geometry. It may be used to represent octahedral atoms with three ligands in a "T" arrangement such as iodine in an iodonium salt. Joining this piece with a linear bond forms a square pyramid, Figure 23B. Joining two octahedral-atom pieces creates an octahedral atom, Figure 23A. An octahedral atom may be used to represent the sp-hybridized carbon of a triple bond or a six coordinate metal with octahedral geometry.

Figure 23. *Atom centers incorporating the octahedral piece*

The assembly and disassembly of the octahedral atom is shown in Figure 24.

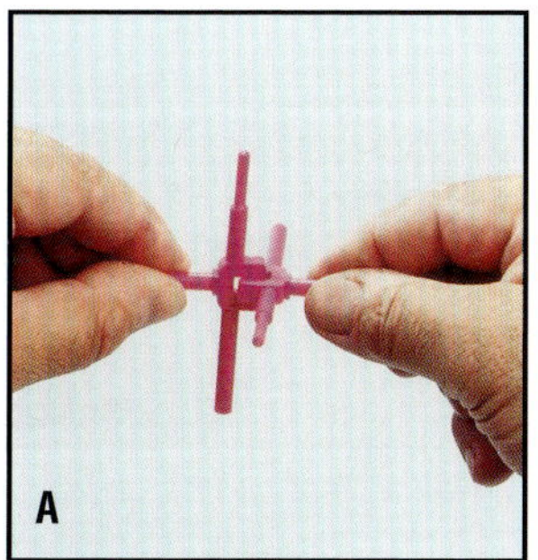

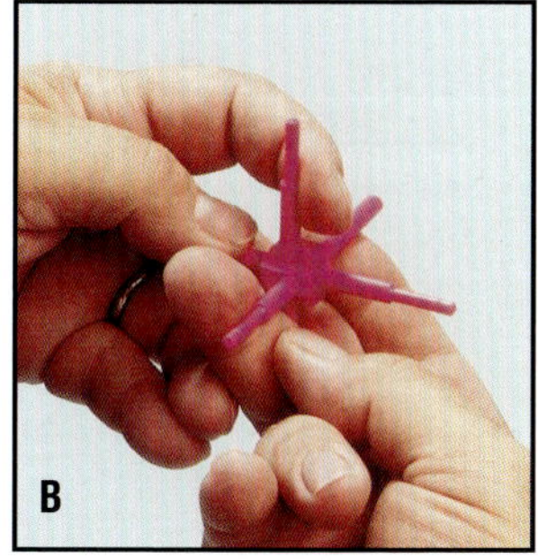

Figure 24. *A) assembly and B) disassembly of octahedral atom center "with bonds".*

LENGTHENING BONDS

The pink bond extender is another sort of wild card. If various colors are available it may be used to mark ligands. It will add 44 pm (0.44Å) to the length of any bond, thus permitting the conversion of the basic pieces into third- or fourth-row elements or lengthening bonds to model partial bonds of transition states. Joining two sp^3 atoms with one bond extender provides a bond of 198 pm (1.98 Å).Two extenders provide a transition state bond of 242 pm (2.42Å). Using one bond extender to join a trigonal or octahedral atom with a tetrahedral atom makes a bond length of 190 pm (1.90 Å). Two trigonal or octahedral atoms joined through one bond extender gives a bond length of 183 pm (1.83 Å) and two extenders increase this to 227 pm (2.27Å).

INSERTING FRAMEWORK ATOMS "WITH BONDS" INTO AN ATOM VISIONS™ SHELL

Figure 25. *The tetrahedral ATOM VISIONS™ hemispheres being placed over the framework atom "with bonds".*

The lower hemisphere with the indented rim is placed over one of the framework bonds. It is then seated with the two bonds at 90 degrees into the rounded slots and held in place with the thumb and forefinger as shown. The other hemisphere is placed over the remaining bond and rotated to match the slots. The ball is formed when the two hemispheres are pinched together.

Figure 26. *The trigonal -octahedral ATOM VISIONS™ hemispheres being placed over the framework atom "with bonds".*

The lower hemisphere with the indented rim is placed over an apical bond. The oval slot is positioned on one of the bonds and the piece held in place with the thumb and forefinger as shown. Add the other hemisphere and align the oval slots. Pinch together to close.

Figure 27. *The trigonal -octahedral ATOM VISIONS™ hemispheres placed over the framework atom "with bonds" pi system of the double bonds.*

When using the ATOM VISIONS™ hemispheres with double bonds, the oval slot is placed over the pi bond. With triple bonds the arrangement is the same and the ball fits into the slots in the pi system.

USING THE ATOM MODEL "WITH BONDS" TO CREATE MOLECULES

You have obtained one of our MOLECULAR VISIONS™ model kits which are available in boxed or bagged form. They all contain pieces which can be joined via the center "U" to form atoms "with bonds". The atoms, in turn, can be joined through the rod and tube bond members to form molecules. An example of this bond making is shown in Figure 28.

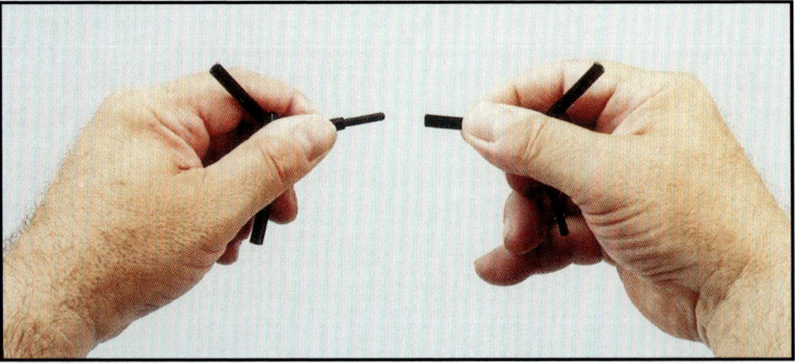

Figure 28. *Joining two tetrahedral atoms by inserting the rod into the tube.*

The pieces should be grasped firmly when pushing the rod into the tube so as not to bend or shear the rod piece. Occasionally the joining offers more friction than desired. This may be alleviated with a light spray of silicone or a thin film of Lubriplate™ lubricant. Oil lubricants do not seem to work as well. All of the other atom centers, "with bonds" may be assembled by joining the rod and tube to form molecules in almost any combination, i.e. to create models for thought.

THREE MEMBERED RINGS AND OTHER STRAINED SYSTEMS

Models of linear (acyclic) compounds and cyclic compounds with rings of five or more atoms are not strained. Four-membered ring compounds (and models) are slightly strained, the angles being about 103° because the ring is puckered and not flat. A three atom ring must be flat: three points define a plane; the 60°-ring angles are constant. To model the three atom ring we use a softer plastic piece, usually colored differently from the other pieces. The sp^3 pieces that are softer are silver-black for carbon, rose for oxygen and turquoise for nitrogen. When forming a three atom ring, a soft piece for the strained bond is inserted into a regular piece for the unstrained bond. The pieces should be supported as shown in Figure 29 to form a symmetrical model ; the rod and tube are then guided together to close the ring.

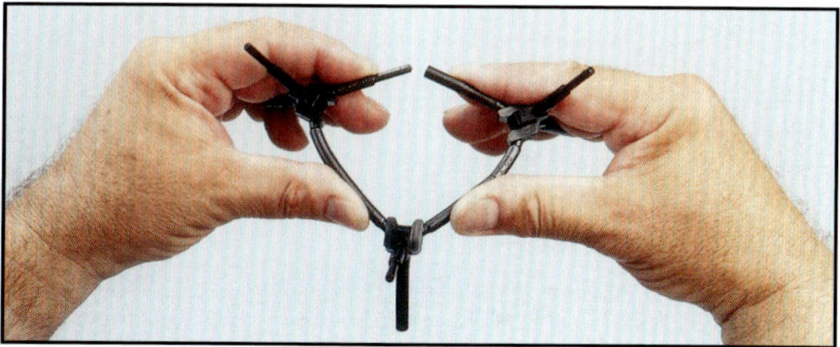

Figure 29. *Forming a three atom ring*

In Figure 30 we can see the models of three atom rings.

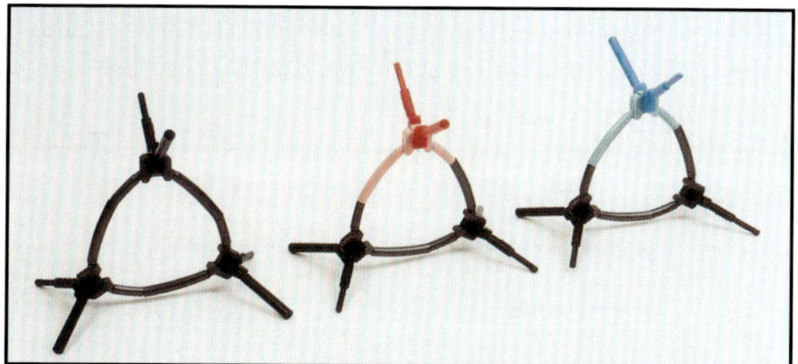

Figure 30. *Models of (L-R) cyclopropane, an epoxide, and an aziridine.*

PROJECTIONS: ACYCLIC AND CYCLIC MOLECULES

Real molecules and models are three dimensional; therefore models of real molecules are also three dimensional; Paper, book pages, and blackboards are not. In drawings then, it is necessary to adopt conventions to represent real molecules in two dimensions. These conventions have been incorporated into various kinds of "projections", e.g. wedge-and-dotted-line, Fisher, Newman, and sawhorse illustrated in Figures 31-41.

A. Wedge and Dotted Line Projections

A three dimensional perspective may be given to drawings of molecules by using "wedges" (—◁ or —◀) and dotted lines (···· or ----- or ▮▮▮▮ or ---▮▮) to represent the out-of-plane bonds. The atom attached to the point of the wedge bond is in the plane; the atom attached to the broad edge is out of the plane towards the viewer. The dotted or dashed lines represent bonds to atoms out of the plane away from the viewer. This convention is illustrated with a model of methane, in Figure 31.

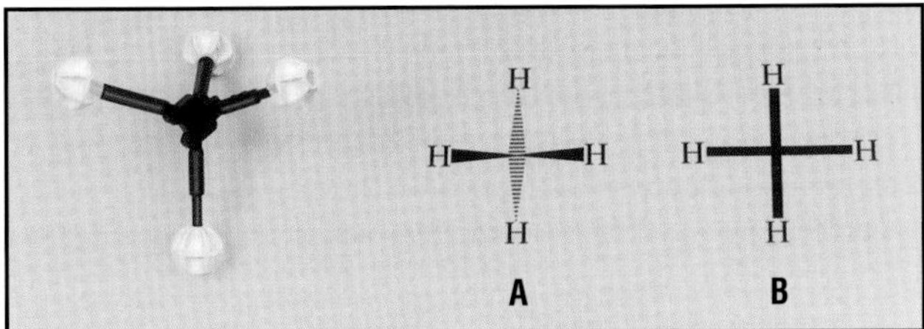

Figure 31. *Methane represented as a wedge-and-dotted-line drawing (A) or Fisher projection (B)*

B. Fisher Projections

In Fisher projections a tetrahedral carbon atom is represented simply by vertical and horizontal lines (which are actually the bonds from that atom). By convention, the horizontal lines represent bonds coming towards the viewer and the vertical lines represent those going away from the viewer, as illustrated for methane in Figure 31 B. In projections of molecules with carbon chains, the carbons of the stem are arranged vertically with the most highly oxidized carbon at the top. Examples are given in Figure 32.

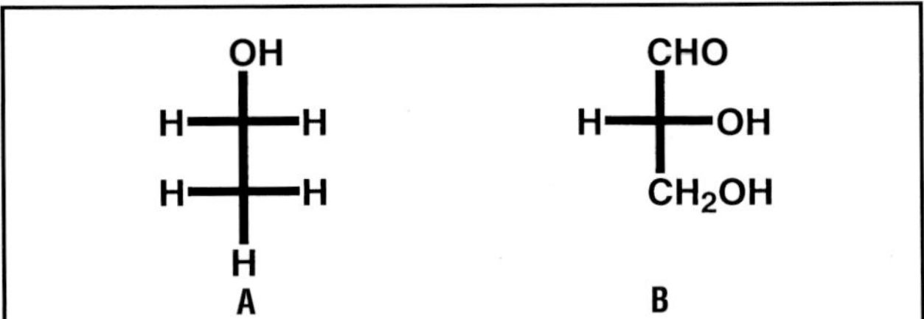

Figure 32. *Fisher projections of CH3CH2OH (ethyl alcohol, A and $CH_2OHCHOHCHO$ (D-glyceraldehyde, B)*

Fisher projections are commonly used because they are simply drawn; but care must be taken to follow the convention. The Fisher projection does not give a picture of the true molecular shape. For example, the projection of glucose does not illustrate the close proximity of the carbonyl carbon (atom 1) to the hydroxyl oxygen on atom 5. "Molecular modeling" Fisher projections, Figure 33 shows how the chain turns back on itself to bring these two atoms into position to form the

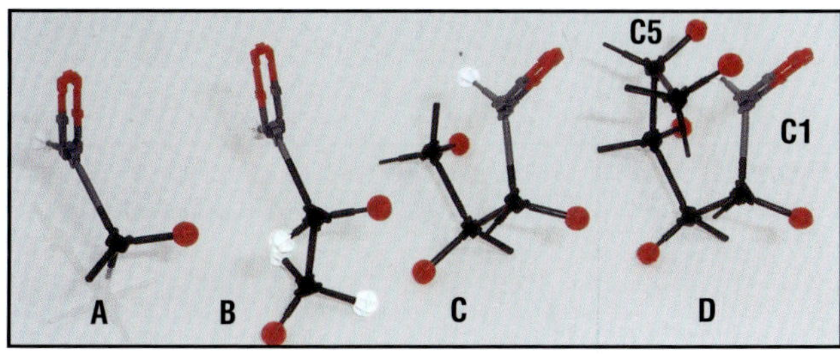

Figure 33. *"Molecular Modeling" Fisher projections.* *$HOCH_2CHO$ (A), $HOCH_2CHOHCHO$ (D-glyceraldehyde B), $HOCH_2[CH(OH)]_2CHO$(C)($HOCH_2CH(OH)CH(OH)CH(OH)CH(OH)CHO$ (D-glucose, D)*

pyranose hemiacetal, Figure 34.

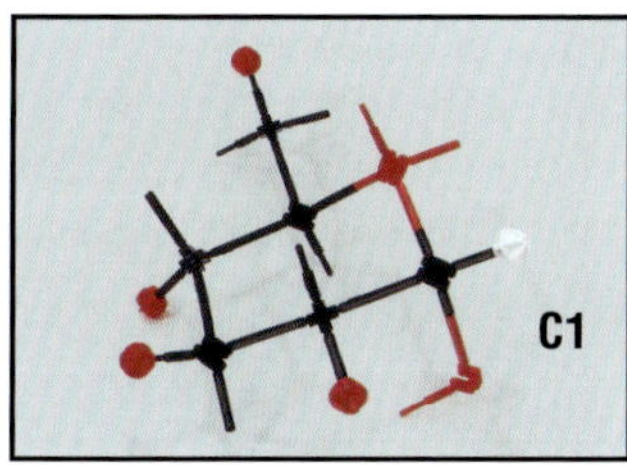

Figure 34. *D-glucopyranose, The tetrahedral oxygen bridges carbon atoms 1 and 5 to form the pyranose ring with the hemiacetal formed at carbon 1*

C. Newman Projections

A Newman projection represents the viewer's "picture" of a molecule looking straight down a bond connecting two atoms. Newman projections of the conformations of 1-propanol are shown in Figure 35. The same view of these conformations

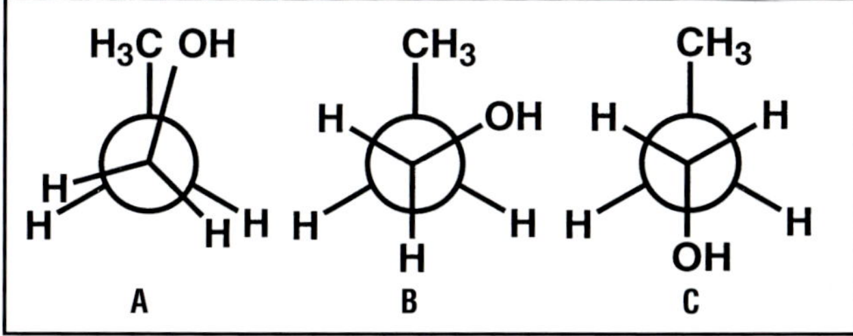

Figure 35. *Newman projections of the conformations of 1-propanol 1-propanol looking down the C1-C2 bond: eclipsed (A), staggered gauche (B), staggered anti (C)*

represented with molecular models is shown in Figure 36.

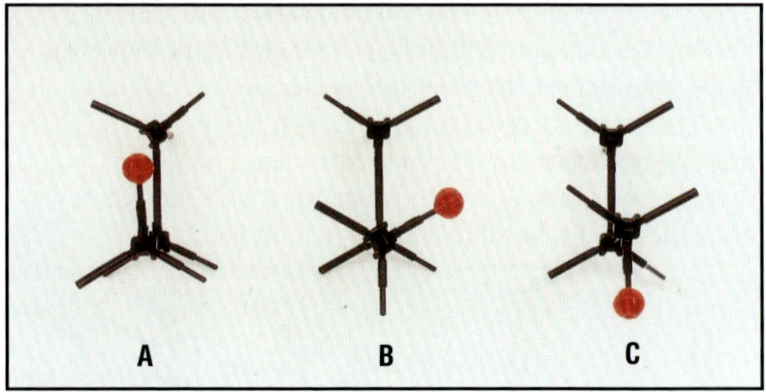

Figure 36. *Molecular model representations of conformations of 1-propanol looking down the C1-C2 bond: eclipsed (A), staggered gauche (B), staggered anti (C)*

Newman projections are very useful in examining molecular conformations.

D. Sawhorse Projections

The sawhorse projection is an oblique view of the bond in the Newman projection. In sawhorse projections, convention dictates that the bond between the joined atoms is drawn diagonally (instead of vertically as in the Fisher projections). The atom attached to the right on the line is behind the atom to the left. Bonds on the right come towards the viewer; bonds on the left go away from the viewer. Figure 37 shows the same conformations of 1-propanol again in sawhorse projections.

CH_3 Back CH_3 CH_3

HO H H H H H OH H H H H

H H Front H OH

A B C

Figure 37. *Sawhorse projections of the 1-propanol conformations, A, eclipsed, B, staggered gauche, C, staggered anti*

Molecular model representations of these sawhorse projections of 1-propanol are shown in, Figure 38.

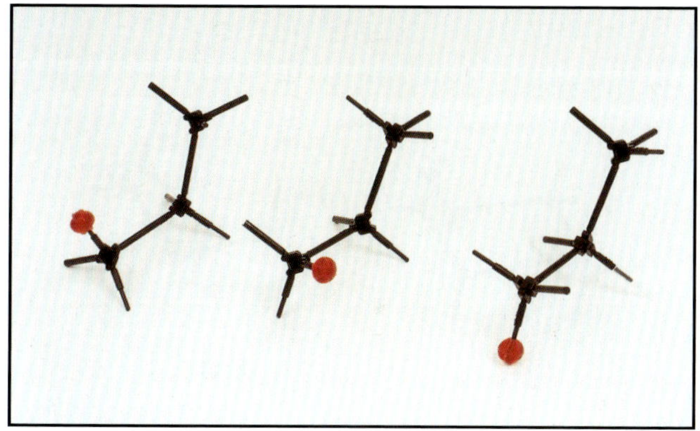

Figure 38. *Molecular model representations of the sawhorse projections of conformations of 1-propanol. Compare with Figure 37.*

All of the previous conventions apply to cyclic as well as acyclic structures. The six carbon cycloalkane, cyclohexane, has been the subject of much investigation. Because of this ring's flexibility, cyclohexane assumes distinct conformations, the two prevalent ones being the boat and the chair. These conformations can be drawn in the Newman projections illustrated in Figure 39 (compare with the corresponding molecular models in Figure 40), or in sawhorse projections, Figure 41.

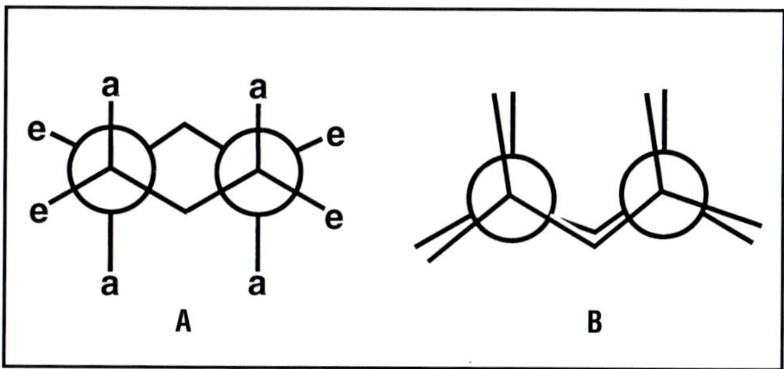

Figure 39. *Newman projections of the chair, A, and boat, B, conformations of cyclohexane.*

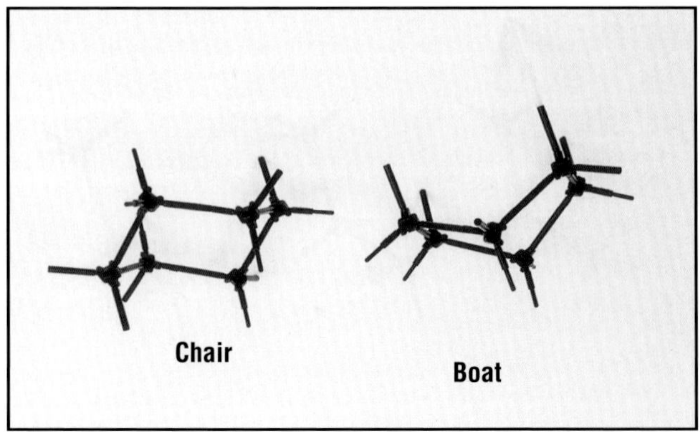

Figure 40. *Molecular Model representations of the Newman projections shown in Figure 39.*

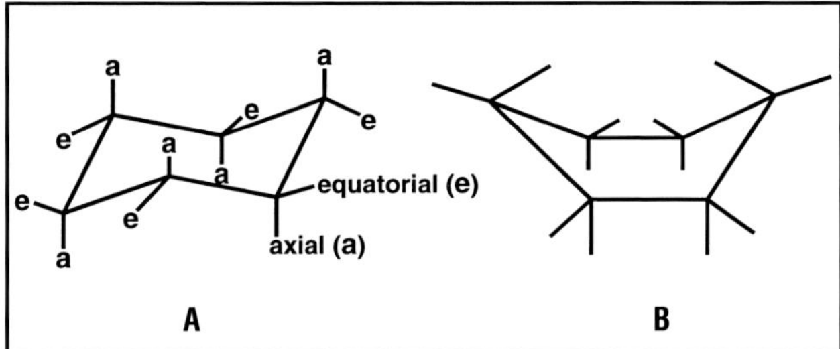

Figure 41. *Sawhorse projections of chair, A, and boat, B, conformations of cyclohexane.*

The chair conformation, which is thermodynamically (energetically) preferred in general, has two sets of six C-H bonds. One set is called "axial"; the other, "equatorial". In Figures 39 and 41 these are denoted by "a" and "e" respectively. In general, substituents (other than a hydrogen atom) prefer an equatorial position which is attained by rotation around the C-C bonds in the ring until a chair conformation is reached in which the substituent is equatorial.

STEREOISOMERS

In addition to using models to examine the connections ("bonds") between atoms and the conformations of molecules, they are especially useful in examining the absolute arrangement of a molecule's component atoms in space, i.e. its stereochemistry. Again, there are conventions used in drawing or describing the absolute stereochemistry of molecules.

Compounds that differ from each other only in the spatial arrangement of their atoms are called stereoisomers. Stereoisomers that cannot be easily interconverted by rotation about a bond are called configurational isomers; those that are easily interconverted by rotation about a bond are called conformational isomers. Certain configurational isomers in the past (and sometimes today) were called *geometric isomers.*

CONFIGURATIONAL-GEOMETRIC ISOMERS

A. Cyclic compounds

Because rotation around bonds within cyclic compounds is restricted, two or more substitutents may be on the same side or different sides of the ring, which leads to configurational-geometric isomers. Cycloalkanes provide good examples. Figure 42 shows examples of two substituents (red and green) on adjacent carbons, i.e. 1-2-disubstitution. The two substituents in 42 (A) are on the same side of the ring; they are cis to each other, one connected to the ring by an equatorial bond, the other by an axial bond. The substituents in 42 (B) are on opposite sides of the ring; they are *trans* to each other, both being connected to the ring by equatorial bonds. The two substituents in 42 (C) are also *trans* to each other, but in a different conformation. These substituents are now both axial, making it much easier to see their *trans* relationship. A comparison of 42 (B) with 42 (C) illustrates the importance of examining all possible conformations of a molecule to determine the most stable one, i.e. the one with the lowest energy. It should be pointed out that while the *cis* structure is a true isomer of the two *trans* structures, the latter two are not isomers of each other but differ only in conformation, which is dependent on energy considerations.

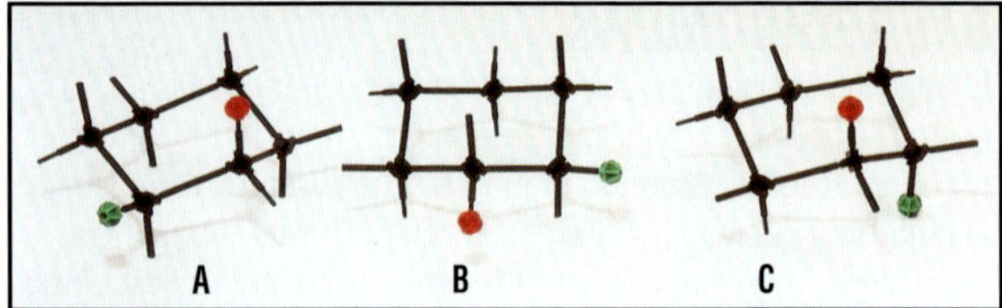

Figure 42. *1,2 substitutions on cyclohexane: cis and trans isomers; conformations. A, cis 1, 2 e, a; B, trans 1, 2 e, e; C, trans 1, 2 a, a.*

Molecular models should be used to explore the 1,3 and 1,4 relationships of two substituents on a cyclohexane ring. A useful drawing shorthand to illustrate geometric isomers of substituted cyclohexanes uses a solid-dotted line notation as shown in Figure 43.

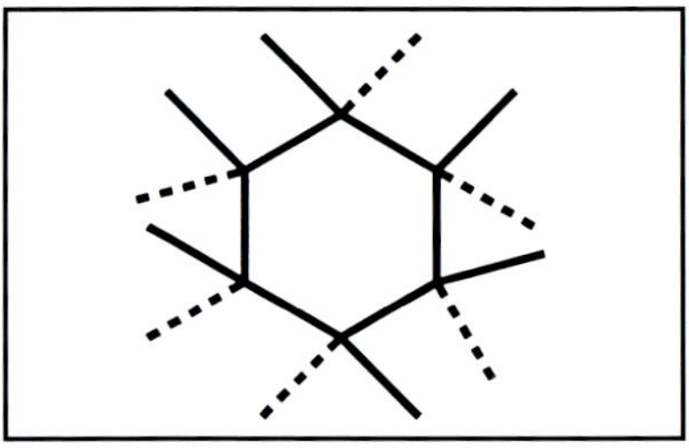

Figure 43. *Representation of Cyclohexane on a plane.*

In Figure 43, the ring is in the plane of the paper. The dotted lines represent bonds going behind the plane (often referred to as alpha) and the solid lines represent bonds going above the plane (referred to as beta). It is easy to see cis and trans relationships from this drawing but spatial relationships such as chair, axial, etc. are not shown and would have to be defined to use this drawing in conformational analysis. We have said that cis substituents which are axial increase the energy (reduce the stability) of cyclohexane more than if they were equatorial. This fact is easily understood if the model is fleshed out with space filling hydrogen Van der Waals radii, Figure 44. The 4-inch yellow balls representing two axial hydrogens make contact with the hydrogens of the methyl group.

Figure 44. *Nonbonded methyl/hydrogen interactions between axial groups on a methylcyclohexane.*

The same phenomenon is illustrated as a "gauche interaction" in, Figure 66.

B. Carbon-Carbon double-bonded compounds

Pi bonding (the "double-bond") also restricts rotation about carbon-carbon bonds and, as described for cyclohexanes above, likewise gives rise to two configurational (geometric) isomers. These isomers are designated *E* (*entgegen:* opposite) and *Z* (*zusammen:* together) which are assigned on the basis of the "priority" given to atoms directly attached to the two double-bonded carbon atoms. The priority system is based on the atomic number of the atom: the larger the atomic number, the higher its priority (with isotopes, the larger the atomic mass, the higher the priority). For example: the atomic numbers of C, O, and N are respectively, 6, 8, 7; thus, the order from highest to lowest priority is O>N>C, And $^2H > ^1H$. This notation is illustrated with 2-amino-3-methoxy-2-butene, Figure 45.

H_2N (7), OCH_3 (8), H_3C (6), CH_3 (6) — **Z** — 7 > 6, 8 > 6

H_2N (7), CH_3 (6), H_3C (6), OCH_3 (8) — **E** — 7 > 6, 8 > 6

Figure 45. *Configurational (geometric) isomers of 2-amino-3-methoxy-2-butene*

In Figure 45, of the substituent atoms directly attached to each double-bonded carbon, N (at. No. 7) has priority over C (at. No. 6), and O (at. No. 8) has priority over C (at. No. 6) respectively. These priority atoms are on the same side ("Z") in A, and on opposite sides ("E") in B. When drawing these structures, the priority atom on each double-bonded carbon should be circled for easy recognition. There are no geometric isomers when the two substituents on either carbon of the double-bond are the same.

C. Metal Centers

Metallic compounds having a square planar or octahedral geometry and two or more different ligands may exist as geometric isomers. For example in the square planar complex of platinum dichloride the chlorine atoms may be attached to the square in adjacent or opposite positions, Figure 46.

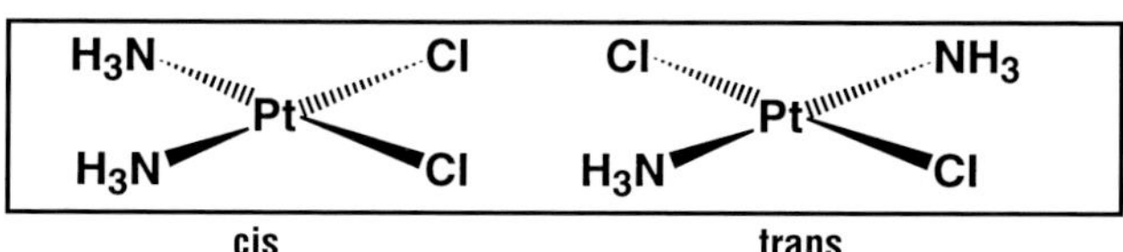

Figure 46. *Geometric isomers of the square-planar platinum dichloride ammonia complex. The isomer with the chlorine atoms adjacent to each other is called the cissomer; the other is the trans isomer.*

Octahedral complexes also exist as *cis* and *trans* isomers. When three identical ligands are arranged in an octahedral geometry, two new isomers are formed. The one with all three ligands in a plane which passes through the metal center, is called the *meridional* isomer, and the other the *facial* isomer. These isomeric octahedral complexes are illustrated in Figure 47.

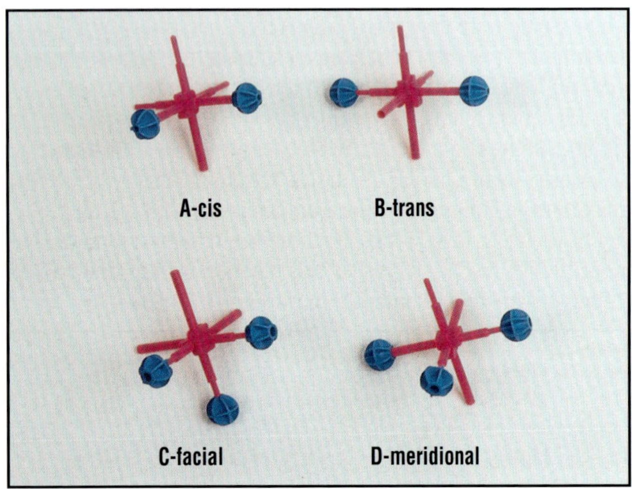

Figure 47. *Geometric octahedral isomers. A is cis; B is trans; C is facial; D is meridional.*

CHIRAL ISOMERS: ENANTIOMERS OPTICAL ISOMERISM

Another form of stereochemistry involving configurational isomers arises from molecular asymmetry. When a structure is devoid of symmetry, it and its mirror-image structure are not superimposable (i.e. identical in all aspects). The structure is said to be chiral, and it and its mirror-image structure are called enantiomers. Both enantiomeric isomers are chiral and they are identical in all chemical and physical aspects except in the "direction" they rotate plane polarized light.

A. Simple aliphatic compounds

An alkane, for example, is composed of only sp^3-carbons and single bonds about which rotation is generally free. If one of these carbon atoms is bonded to *four* different groups, the molecule is asymmetric, i.e. chiral, and it and its mirror-image structure are not superimposable; they are enantiomers. The absolute arrangement of the four groups bonded to this carbon atom of an enantiomer is called its *chirality* and is designated R (rectus) or S (sinister). These designations are based on the "priority" of the atoms attached to the chiral carbon, the priorities again reflecting atomic number. Since there are four atoms to arrange, the priorities are 1 for the atom of highest atomic number, down to 4 for the atom of lowest atomic number, Figure 48.

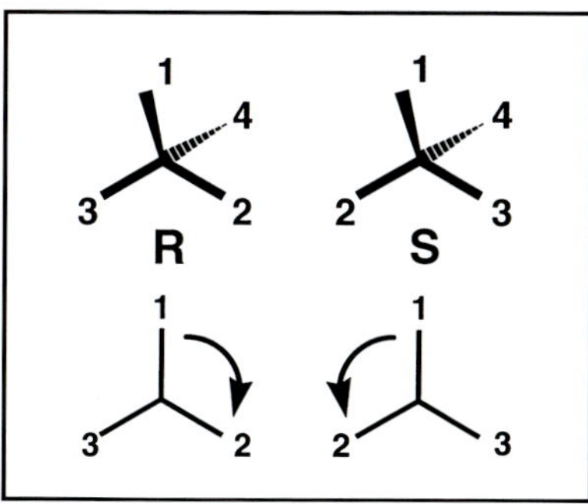

Figure 48. *Example of absolute chirality: R and S configurations at a chiral center, when priority 4 is behind the plane.*

An sp^3-carbon atom having two identical substituents is symmetrical, i.e., achiral; it and its mirror image are superimposable (identical) and isomerism of this type is not possible.

B. Metal centers

The isolation of compounds containing covalently bonded metal ligands has led to the discovery of a different type of geometric isomerism. While arrangements of monodentate ligand (for example: Cl, HOH, NH_3) around octahedral atoms also lead to stereochemical isomers, they are rare compared to those associated with tetrahedral atoms. Bidentate and tridentate ligands form stable compounds which may exist as geometric and optical isomers. Ethylenediamine is a bidentate ligand which forms stable tris complexes with octahedral atoms. The length of the carbon chain between the two NH_2 groups restricts each ligand to a cis arrangement. Three such ligands may be positioned around the octahedron resulting in a unique spatial arrangement to provide chirality, which may be demonstrated by constructing both enantiomers. In this representation only one sp^3 piece per atom is used for simplicity (Figure 49).

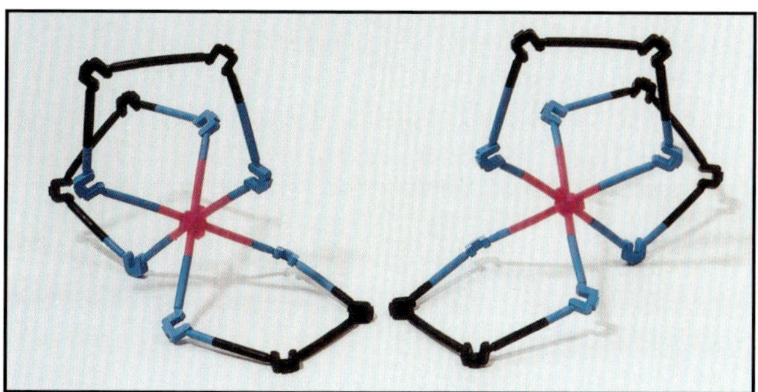

Figure 49. *Optical isomers of a tris-ethylenediamine complex*

C. Chiral molecules without sp^3 chiral ATOMS

There are several classes of chiral molecules whose chirality arises from asymmetry other than that associated with an sp^3 chiral atom. The helical protein chain ("alpha-helix", Figure 74), which resembles a spiral staircase, is itself chiral...a chirality in addition to that arising from its asymmetric atoms. Because such helices may coil in a clockwise as well as counter-clockwise spiral, these two forms are themselves asymmetric (nonsuperimposable) mirror images. The two forms exhibit the same optical activity but of opposite rotational sign.

An example of another class of such chiral compounds is trans-cyclooctene which exists as two nonsuperimposable optical isomers, Figure 50.

Figure 50. *A chiral R trans-cyclooctene*

A third class of chiral molecule consists of 1,3-substituted 1,2-propadienes, C=C=C, called allenes (Figure 16G). Again, as opposed to chirality arising from an sp^3-carbon atom with four different substituents, chirality here requires only that C-1 has two different substituents and C-3 has two different substituents, even though the two substituents on C-1 can be the same as those on C-3. These molecules are asymmetric because the plane of C-1 with its substituents is always perpendicular to the plane of C-3 with its substituents, which makes them nonsuperimposable with their mirror images. In a sense, then, C-2 is a chiral center effecting molecular chirality. An example is 1,2-propadiene, Figure 51.

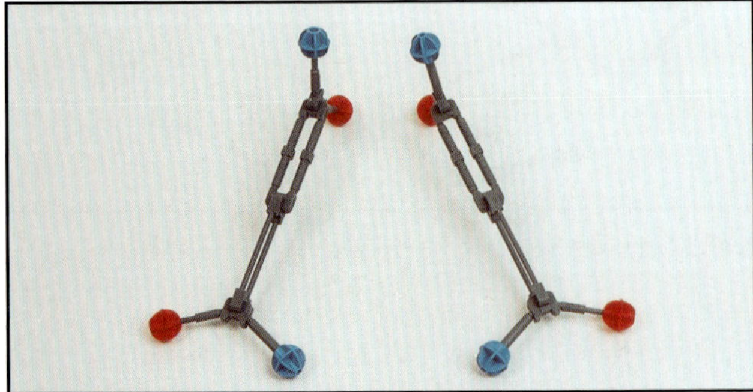

Figure 51. *Chiral allenes.*

MODELS TO INVESTIGATE REACTIONS

REACTIONS

What has generally been omitted from suggestions for model usage is the investigation of reactions. The objective in this quest is to help account for all the atoms in the starting materials. One is encouraged to make models of all starting materials and products and to make sure that the products contain all the atoms of the starting materials. The reaction of bromine with E 3-methyl-2-pentene is an illustration (Figure 52).

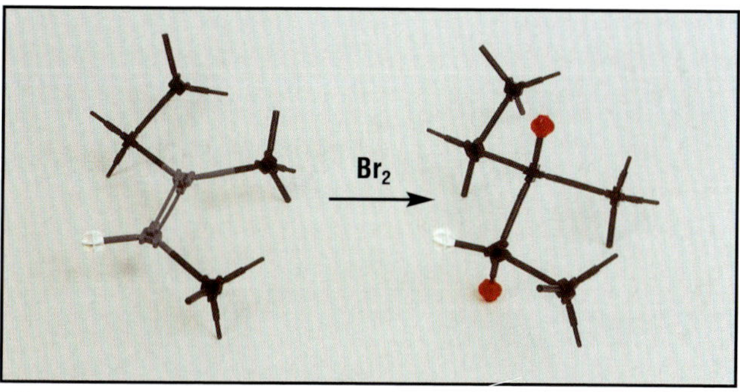

***Figure 52**. Investigation of a reaction: bromination of an alkene, E 3-methyl-2-pentene*

The reaction in Figure 52 involved a symmetrical reagent Br_2. When the reagent is unsymmetrical, such as HBr, and the alkene is also unsymmetrical, two products are possible although usually one predominates. The carbon atom of the potential carbocation and the carbon atom being protonated are frequently reversed by students contemplating such reactions. Molecular models can be helpful by showing the pi complex with a hydrogen marker attached to a pi-bond connector, Figure 53.

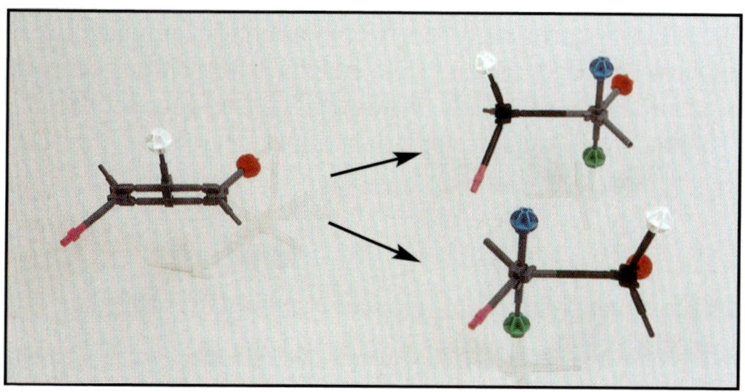

***Figure 53.** Models of a protonated alkene; carbocation intermediates.*

It should be noted that a bond extender is also used as a marker in this illustration. The intermediate carbocations are modeled at the right to show that the hydrogen marker and carbocation (trigonal-bipyramid atom with green and blue markers for the empty "p" orbital) are on adjacent atoms.

Another reaction illustrating this use of molecular models is the Ozonolysis of alkenes. In this reaction the double bond is cleaved, leading to the formation of two new molecules both of which contain a carbonyl function. The double-bond-pi piece is constructed from the half-double-bond pieces which permits the user to separate the double bond in to two pieces to simulate the breaking of this bond. The half-double bonds may be capped with additional half-double bond pieces to represent the newly formed carbonyl compounds (Figure 54).

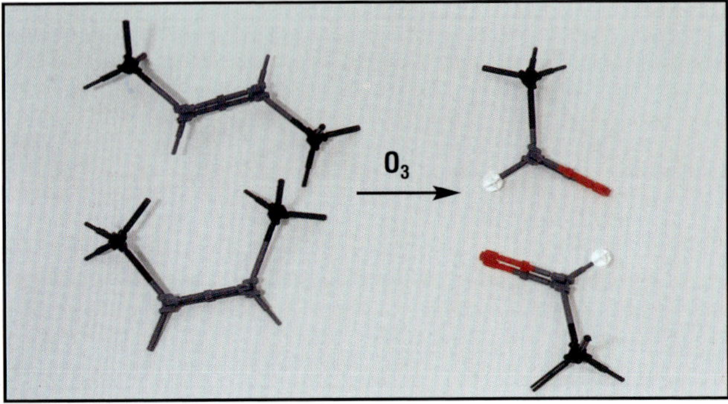

Figure 54. *Ozonolysis of an alkene*

Conversely, if the starting alkene is to be predicted from the products of ozonolysis(e.g. in an identification study) the two half-double bonds are joined to form the alkene. The geometric isomerism of the original alkene is not transferred to the products; i.e. the E and Z isomers of an alkene give the same products.

RETROSYNTHESIS

Molecular models are also useful in planning multistep syntheses from the reverse aspect, called "retrosynthesis". In this methodology the product structure is initially examined for functional groups. One bond is separated to examine a possible one-step reaction which will remake that bond. The Grignard synthesis of 2,3-dimethyl-3-pentanol, for example, may be examined this way (Figure 55).

1) A model of the product is constructed (A).

2) The product's functional group is examined. It is a tertiary alcohol.

3) The method of synthesis to be explored is stated: A Grignard reaction. This reaction is defined: The addition of an organo magnesium compound to an aldehyde or ketone (or sometimes an ester).

4) It is realized that the Grignard reaction would yield a new carbon-carbon bond, i.e., from the carbon atom of the Grignard compound to the alpha carbon of the product alcohol.

5) This bond-forming step is reversed by separating a C-C bond of C-C-OH group of the product model (B). It should be noted that there are three bonds which could be separated and any one of them is valid for consideration if the required starting materials are available.

6) Model pieces are selected to make the necessary functional groups on the two pieces that "will undergo" the Grignard reaction to give the alcohol (C).

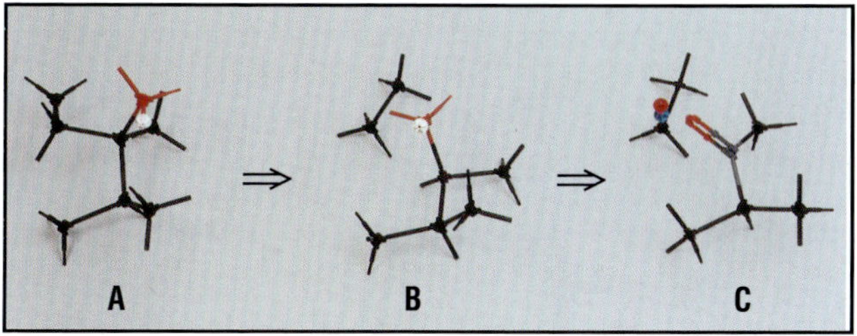

Figure 55. *Retrosynthesis: dissection of a tertiary alcohol*

The stepwise approach to reactions with the use of models will help to prevent common mistakes e.g. transposing the oxygen atom as shown in Figure 56.

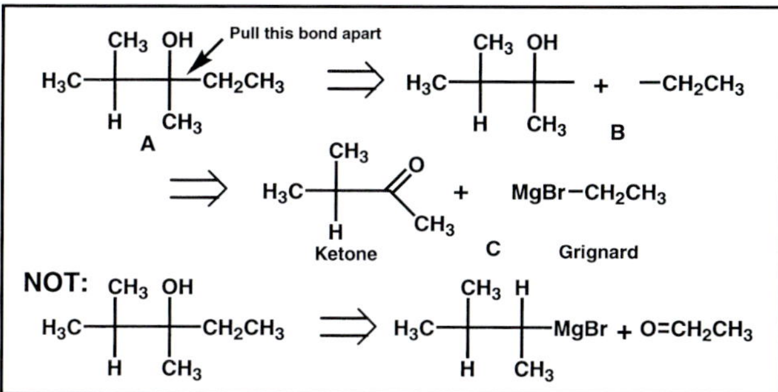

Figure 56. *Correct and incorrect disconnections in a retrosynthesis.*

The same procedure may be useful for analyzing all bond making steps in reactions such as the Williamson ether synthesis, aldol condensations, Diels-Alder cycloadditions, alkylations, etc.

USING MODELS TO INTERPRET SPECTRA

When making assignments from spectral correlation charts it is important to consider the connectivities between atoms/groups which could explain a puzzling chemical shift or proton coupling in the NMR spectrum. For example, the NMR and IR spectra of ethyl benzoate are similar to those of phenyl propionate. It can even be difficult to differentiate between propiophenone and 4-methylacetophenone based solely on an empirical formula and an NMR spectrum if proton coupling is not adequately considered, which often happens when retrieved information is not carefully retained.

Molecular models can assist. For example, an "unknown" aromatic compound possesses an alkyl group and a carbonyl. If its NMR spectrum shows a coupled triplet and quartet indicative of an ethyl group, then this fragment is constructed and one notes that there is only one bond left for attachment to the whole molecule. A signal in the aromatic region which integrates for five protons indicates that this is a simple phenyl ring with only one point of attachment. Carbonyl absorption in the IR spectrum indicates a fragment with two bonds for attachment (Figure 57).

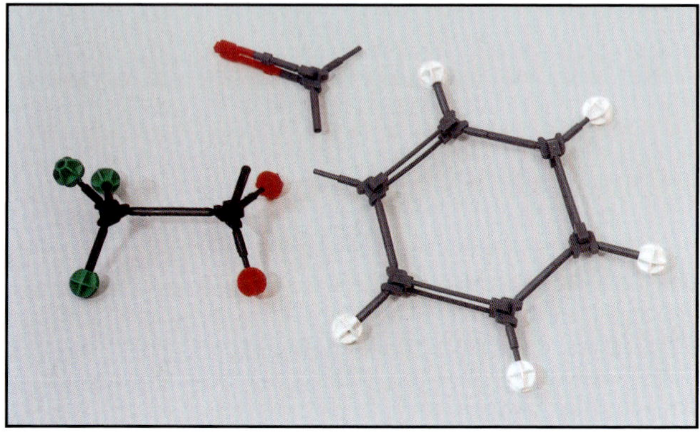

Figure 57. *Molecular fragments determined from spectral interpretations*

The logic is that the phenyl and ethyl groups each must be attached to the carbonyl; if the ethyl were attached to the phenyl there is no point of attachment for the carbonyl. When the test molecule is assembled it is necessary to work backwards to be sure that this molecule would be responsible for the observed spectral data. The ethyl group should be associated with the quartet and triplet in the NMR spectrum. In this case the quartet appears in the NMR spectrum at 3.5 ppm. The model indicates that the methylene is directly attached to the carbonyl; spectral-correlation charts indicate that such a methylene should resonate close to $\partial 2$. If the empirical formula is given as $C_9H_{10}0_2$, the apparent inconsistency may remind the user that an atom is missing in the model and that placement of an oxygen atom between the ethyl and carbonyl groups would result in a structure (ethyl benzoate) consistent with the data.

MODEL BUILDING FOR ENHANCED VISUALIZATION

ATOM VISIONS™ HEMISPHERES TO ENLARGE THE ATOM CENTER

The ATOM VISIONS™ hemispheres may be added to any structure for enhanced visualization of the atoms. Some examples are given below.

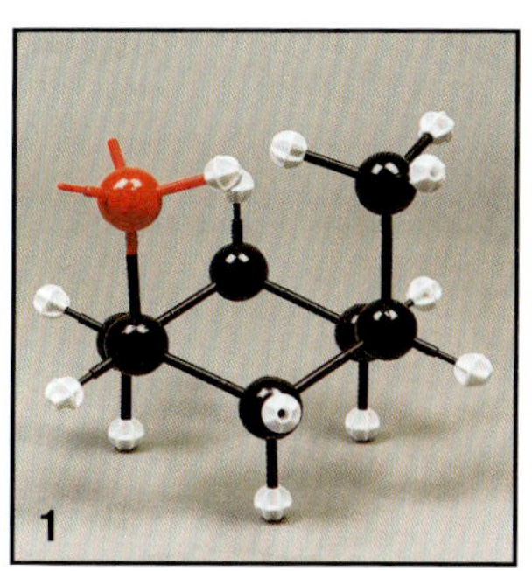

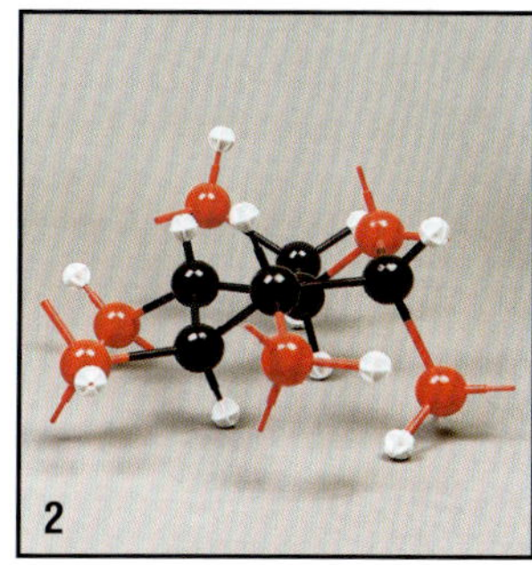

Figure 58. *ATOM VISIONS™ enhanced models for 1) (IS,3R)-3-methylcyclohexanol-1 2) Glucose*

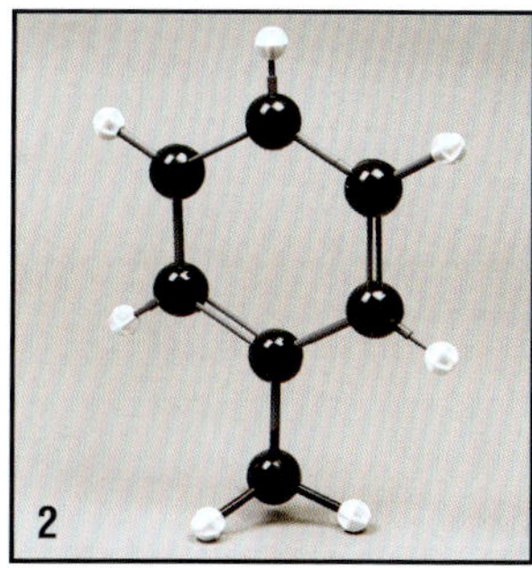

Figure 59. *ATOM VISIONS™ enhanced models for 1) Glycine 2) Toluene*

USE OF COLORED MARKER BALLS IN MODELING

A. Markers for differentiating "like" atoms.

Colors play a very important role in differentiating similar things. With molecular models, different colored marker balls may be used to represent different hydrogen atoms, thereby differentiating among otherwise "similar" hydrogen atoms in different environments, Figure 60.

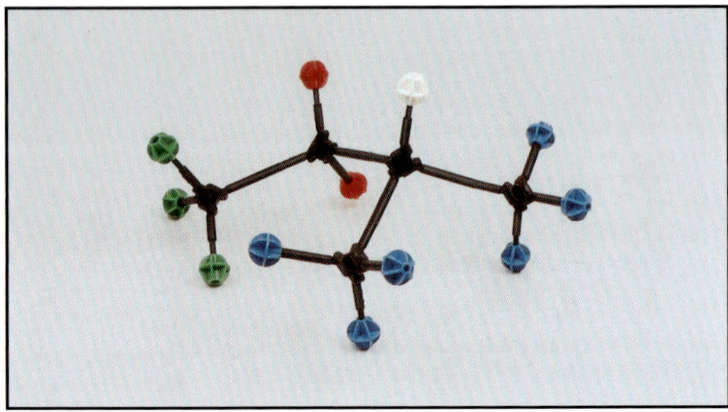

Figure 60. *Colored marker balls used to mark "equivalent" and "non-equivalent" hydrogens.*

B. Orbital symmetry

By adding green and blue marker balls or 3-in. soft foam balls to the trigonal bipyramid, a "color coded" representation of "p" orbitals may be made to illustrate orbital symmetry (Figure 61).

Figure 61. *Models of orbital symmetry*

The aromatic pi system may also be modeled to show a circle composed of orbitals, Figure 62.

Figure 62. *Modeling aromatic systems*

1. Electrocyclic stereochemistry

Two-toned trigonal bipyramid "atoms" may also be used to determine the stereochemistry of ring opening or closing in electrocyclic reactions. The bond involved in the reaction is represented by the linear piece of a trigonal bipyramid and is labeled with marker balls to show symmetry, Figure 63.

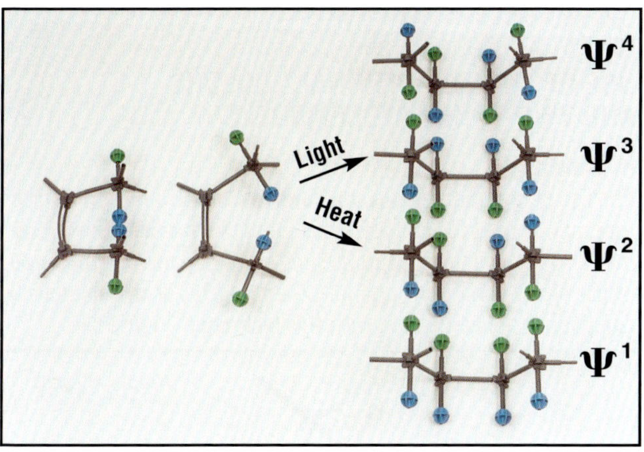

Figure 63. *Use of color-coded models to determine the course of an electrocyclic reaction.*

Breaking the bond produces two "p" orbitals which are rotated to form the frontier orbitals of the product. The rotation to generate the HOMO or LUMO state produces the required con- or disrotatory motion.

2. Sigmatropic rearrangements

Similarly, sigmatropic rearrangements can be illustrated by the opening of a vinylcyclopropane. The strained bonds should be constructed of the softer atom pieces, silver-black for sp^3 and black trigonal and linear pieces for the trigonal bipyramid, Figure 64.

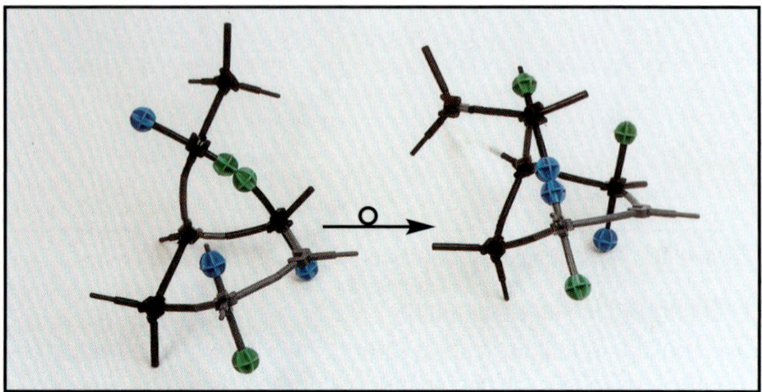

Figure 64. *Modeling a sigmatropic rearrangement.*

When the bond is broken, it should be noticed that bonding at the other end of the allylic system requires inversion of the migrating carbon and results in the methyl group being correctly placed.

USE OF "NONTRADITIONAL" COLORS AND TWO-TONED ATOM MODELS

A. Markers

The use of nontraditional colors for some of the atoms in a structure may be used to emphasize that atom. Such color distinctions may also be advantageous in keeping track of parts of a molecule or particular bonds. For example, the bridgeheads in polycyclic compounds are easily seen with this technique, Figure 65.

Figure 65. *Marking the bridgeheads with colored atoms.*

B. Conformations

The chair and boat conformations of cyclohexane represent classic ring structures, which are easily constructed and interconverted using molecular models. Two-toned tetrahedral "atoms", made from a red and a black sp^3 piece, may be used to illustrate the gauche and anti conformations of cyclohexane substituted axially and equitorially with methyl, Figure 66 (A and B) or to differentiate between the axial and equatorial bonds or *cis* and *trans* related bonds (C and D).

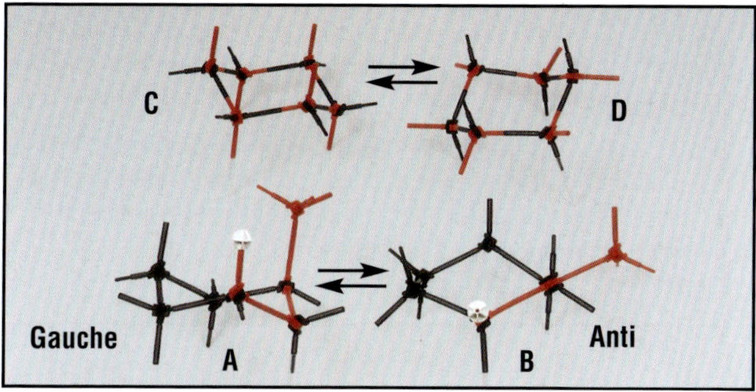

Figure 66. *Models showing the use of two-colored atoms.*

RESONANCE MODELS

The half-double-bond-pi piece serves to cap the ends of a pi resonance hybrid system when used in a combination with trigonal atoms and sp^2 pieces. The trigonal atoms represent the continuous sigma system and the half-double-bond-pi and sp^2 pieces represent the continuous pi system. A resonance-hybrid model of an amide is illustrated in Figure 67.

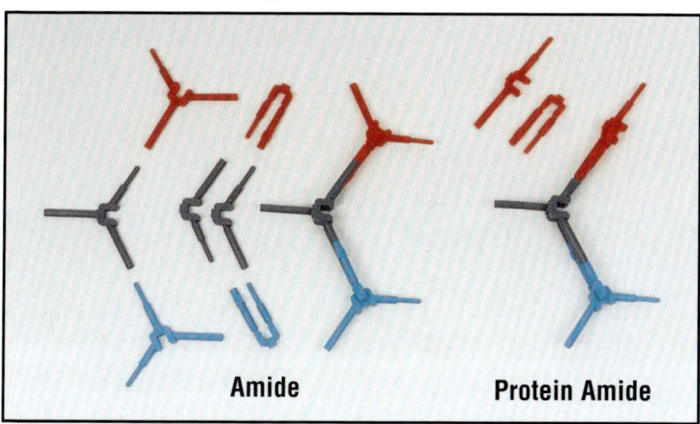

Figure 67. *Construction of a resonance hybrid model of an amide*

With these pieces in different colors, models of allylic-type resonance systems such as allylic and enolic ions and free radicals, enamines, amides, carboxylate ions, and nitro groups can be constructed. If trigonal pieces are used in place of sp^2 pieces, a resonance model of trimethylene methane, carbonate ion, or nitrate ion may be constructed. The "resonance model" may be extended to illustrate other systems such as an enolate of a beta-diketone for use as a ligand.

METAL LIGANDS

The cube in the center of the double bond made from half-double-bond-pi pieces, permits its attachment as a ligand to other atoms, e.g., as a representation of the Wilkinson catalyst bonded to an alkene (Figure 68). The pi bond connector is used to attach the ligand. The cube in the center of one of the pi bonds of the triple-bonded atoms also allows attachment of this piece as a ligand.

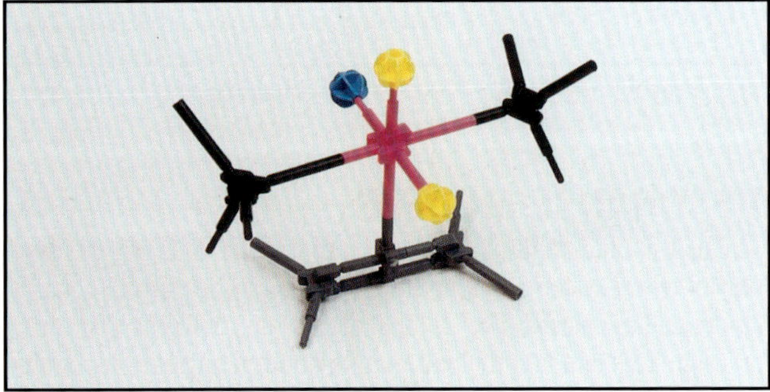

Figure 68. *Model of a pi bonded ligand.*

Through the use of the various bonding pieces shown in Figure 11, a wide variety of metal ligands can be modeled, Figure 69.

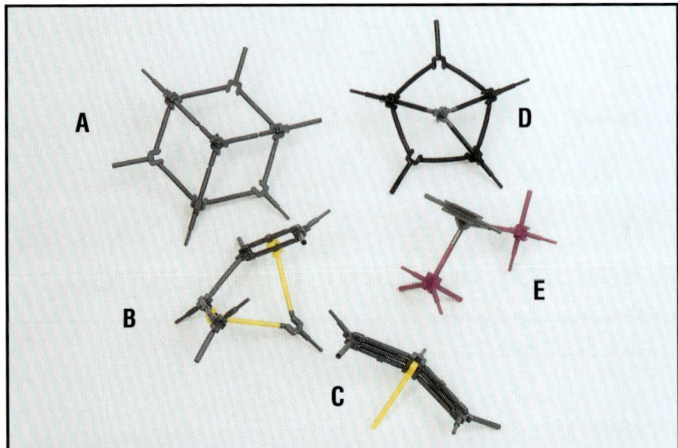

Figure 69. *Modeling organometallic ligands. A. a benzene ligand, B. a tetrahapto ligand, C. an allyl ligand, D. a cyclopentadienyl ligand, E. a biscobalt alkyne complex*

The cyclopentadienyl ligand is attached to a tetrahedral atom (titanocene) or an octahedral atom to form ferrocene, Figure 70.

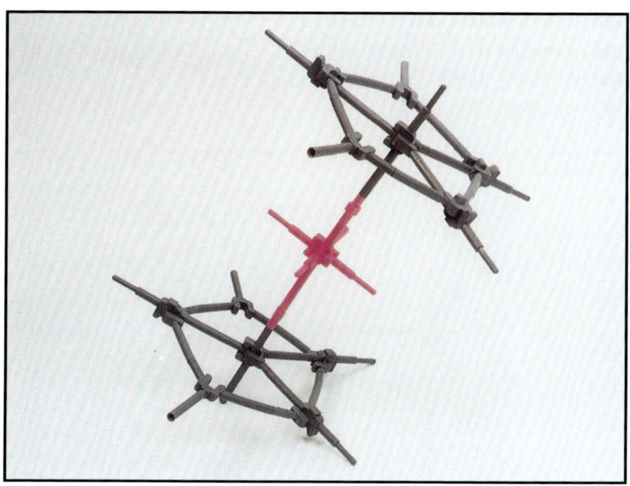

Figure 70. *A model of ferrocene.*

Other ligands not requiring special bonding pieces are easily constructed from model parts. An example is the porphine ligand, Figure 71.

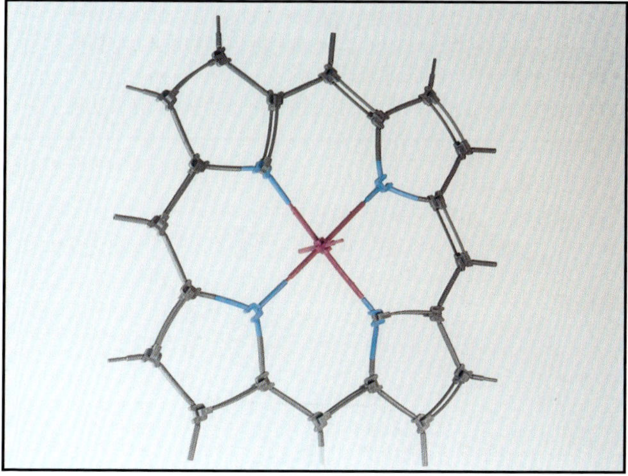

Figure 71. *A porphine ligand.*

HYDROGEN BONDS

Hydrogen bonding involving H_2O is modeled by inserting the tube end of a bond extender into a marker ball and inserting the rod of the donor into the small hole of the same ball. The rod of the bond extender is then inserted into the tube of an acceptor H_2O molecule, Figure 72.

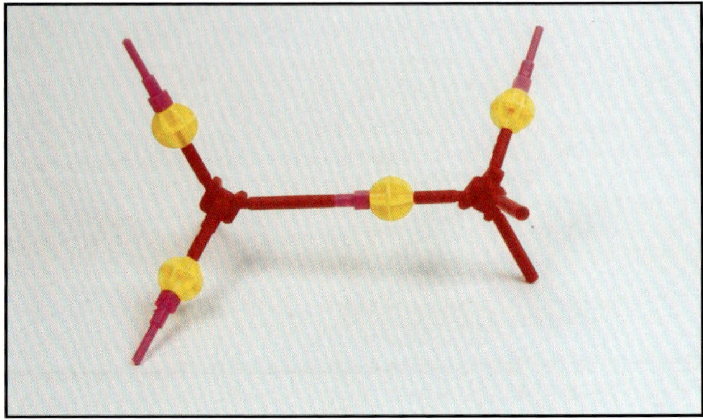

Figure 72. *Modeling hydrogen bonding involving H_2O.*

The linear bond provides a rigid hydrogen bond of the correct van der Waals radii between a donor and acceptor such as found in base pairs (Figure 73) or peptides such as an alpha-helix (Figure 74). The hydrogen bonds are yellow. When used in groups of three, the bond extenders similarly form a hydrogen bond, as illustrated by one of the hydrogen bonds in Figure 73.

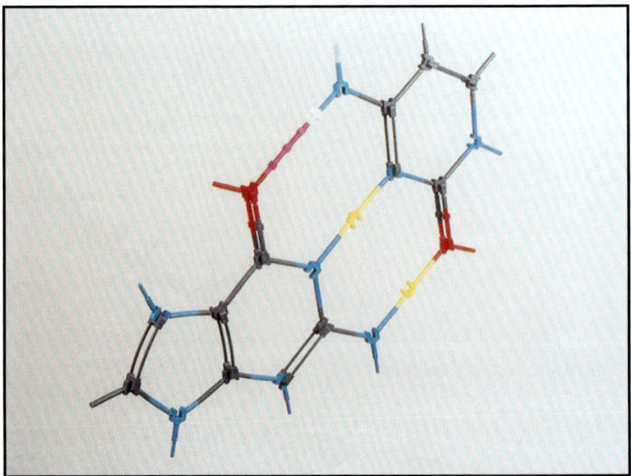

Figure 73. *A base pair*

Figure 74. *Protein Alpha-helix*

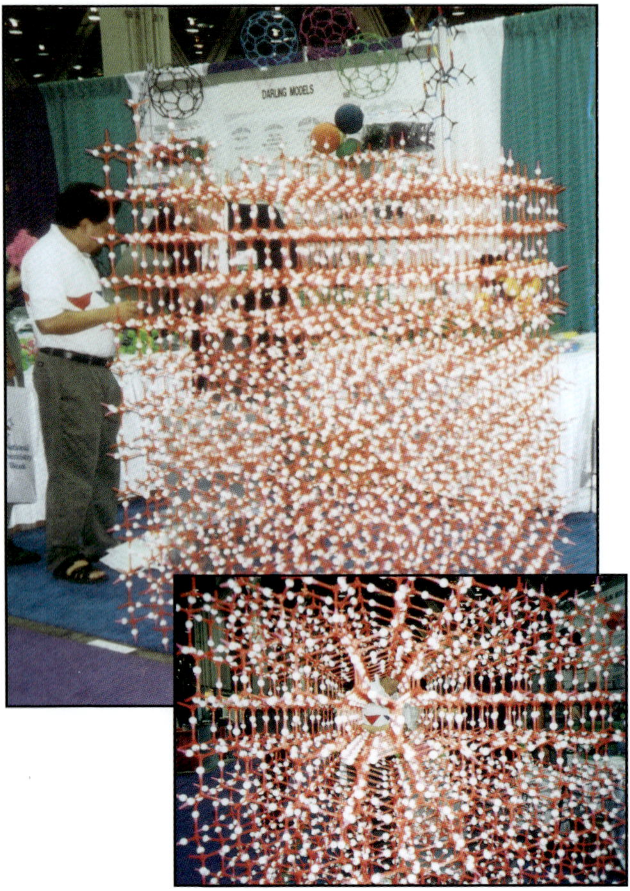

Figure 75. *A Molecular Visions™ model of an ice crystal lattice.*

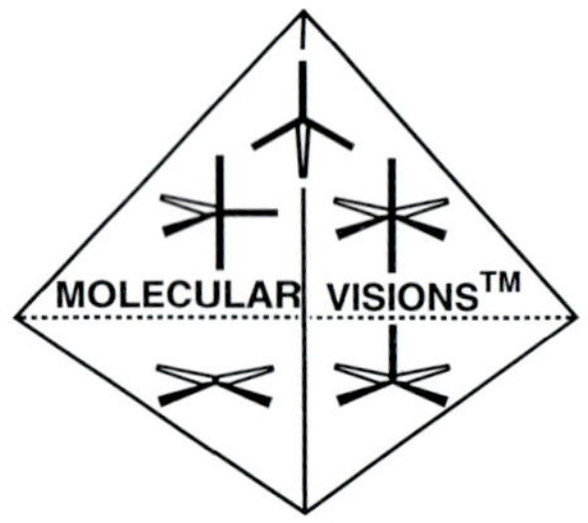

DARLING MODELS INC.
P. O. BOX 1818
STOW, OH 44224

ISBN 978-0-9648837-03

VOICE: 330-688-2080 • FAX: 330-688-5750
E-MAIL: darling@darlingmodels.com
WEB SITE: www.molecularvisions.com
or: www.darlingmodels.com